*Dieses Buch ist gewidmet allen Förderern der „Alexander von Humboldt", die den in der Geschichte der Seefahrt einmaligen Umbau eines Feuerschiffes zum Rahsegler ermöglichten, insbesondere den zahlreichen Helfern, die in ihrer Freizeit viele Arbeitsstunden an Bord des Schiffes geleistet haben.*

*This book is dedicated to all sponsors and promoters of the "Alexander von Humboldt" who made this historically unique conversion of a lightship to a squarerigger possible, in particular to the countless helpers who worked so hard on the vessel in their spare time.*

Clas Broder Hansen
Hans Jürgen Hansen

# Alexander von Humboldt

Vom Feuerschiff zum Windjammer für die Jugend
A Lightship Becomes a Windjammer for Young People

Photos:
Günter Barty, Leo R. Beckmann, Herbert H. Böhm,
Axel Olaf Gäde, Andrea Grammes, Kai Greiser,
Clas Broder Hansen, Hans Jürgen Hansen,
Monika Kludas, Hero Lang, Bodo Mueller, Renard,
Bernd Schulz, Ino Steffen

Schiffspläne / Plans:
Zygmunt Choreń, Peter Knuth

Geleitwort / Foreword:
Kapitän zur See a. D. Hans Freiherr von Stackelberg
Präsident der Sail Training Association Germany

Urbes Verlag

Alexander von Humboldt (1769–1859), Forschungsreisender, Geograph, Diplomat und Mitbegründer der Berliner Universität, auf dessen Namen der Windjammer für die Jugend getauft wurde, am Orinoko in Südamerika. Zeitgenössischer Stich von Johann Joseph Freidhoff in der Messe des Schiffes nach dem Gemälde des Berliner Akademiedirektors Friedrich Georg Weitsch aus dem Jahre 1806.

Alexander von Humboldt (1769–1859), explorer, geographer, diplomat and co-founder of the University of Berlin, with whose name the "windjammer for young people" was christened, by the Orinoco in South America. Contemporary engraving by Johann Joseph Freidhoff hanging in the chip's mess, after the painting by the Director of the Berlin Academy, Friedrich Georg Weitsch, dating from the year 1806.

## *Zum Geleit*

*Eine Idee wurde durch das nahtlose Zusammenspiel verständnisvoller Behörden, großzügiger Stifter und Spender sowie den selbstlosen Einsatz vieler Idealisten zur Wirklichkeit: Ein Windjammer für die Jugend; realisiert mit einem durch seinen idealen Rumpf dafür prädestinierten Feuerschiff!*
*Warum aber im Zeitalter der Technik ausgerechnet ein Segelschiff, ein rahgetakelter Großsegler, der weder als Ausbildungsinstrument der Marine, noch zur Nutzung durch die Handelsschiffahrt, oder gar als geldbringender Kreuzfahrtsegler, sondern für junge Menschen gedacht ist, die in den meisten Fällen später sicher gar keinen seemännischen Beruf ergreifen werden? Die Überlegungen dazu waren vielschichtig: Unsere schnellebige Zeit, geplagt von Konkurrenzkämpfen und egoistischem Karrieredenken, läßt allzu viele menschliche Werte auf der Strecke bleiben; Werte wie Toleranz, Kameradschaft und Einsatzbereitschaft für den Mitmenschen, aber auch Mut, Pflichtbewußtsein und Selbstüberwindung, die ein segelgetriebenes Fahrzeug von seiner Crew fordert und somit zwangsläufig wieder aufleben läßt. Die Natur trifft plötzlich hautnah auf den hier besonders aufnahmebereiten Menschen und fördert in seltener Intensität sein Umweltbewußtsein.*
*An Bord eines Segelschiffes fehlt die so viel Sicherheit suggerierende Glasscheibe zwischen dem Element und dem Menschen, und körperliche Arbeit an Deck und in der Takelage bei Wetter und Wind sorgt für eine sinnvolle Abhärtung, wobei die Härtetage als Bewährung bei Seegang und Sturm von Petrus gemacht werden und die Schiffsführung wie das Decksvolk in gleichem Maße treffen. Das Erkennen von Wetterentwicklungen und der Möglichkeit ungeahnter Kräfteentfaltung aufgewühlter Seen, wenn das grüne Wasser anstelle von Gischt an Deck schlägt, wird sicherlich auch nirgendwo so körpernah erlebt wie auf einem von seiner Takelage abhängigen Segler. Das aber verlangt wiederum von jedem Einzelnen, Teil eines Teams zu sein, unter fordernden Verhältnissen auf seine Umwelt Rücksicht zu nehmen, Verantwortung für sich selbst, für andere und für das Schiff zu tragen und nicht zuletzt Selbstdisziplin zu üben.*
*Das deutlich werdende Leitmotiv für jede Segelschiffsbesatzung: „Nur gemeinsames Zupacken bringt den Erfolg", führt vom Erfassen des Teamworkgedankens mit seiner Notwendigkeit für die Praxis zum Erkennen des Wertes gemeinschaftlicher*

## *Foreword*

*An idea was born, and thanks to the integrated teamwork of understanding officials and generous sponsors together with the selfless effort of large numbers of idealists, this idea became realitiy: a windjammer for young people, realised with a light-vessel whose ideal hull predestined her for this purpose.*
*But why, in the age of technology, should it have to be a sailing ship, a square-rigged tall ship, intended not as a training instrument for the Navy, not for use by the merchant service, not even as a money-spinning cruising yacht, but for young people, most of whom will certainly not be seeking employment in a nautical trade in the future? There were many reasons pointing in this direction. Our fast-moving age, plagued by rivalry and egoistic career considerations, leaves far too many human values by the wayside; values such as tolerance, cameraderie and readiness to assist fellow human beings, as well as courage, a sense of duty and willpower, all values which a sailing vessel demands of its crew and thus necessarily brings to the fore again. Natur suddenly appears face to face with the people at sea, who are particularly receptive in this situation, and strengthens their environmental consciousness with a rare intensity. That pane of glass between man and the elements which suggests so much security on land is not there on board a sailing vessel, and physical work on deck and in the rigging during wind and weather provides for a certain toughness, whereby the days of hardship are ordained by St. Peter, the patron saint of weather, as a proving ground in storm and swell and affect the officers and the deckhands equally. The recognition of weather developments and the potential of the unimagined unleashing of forces displayed by turbulent seas when the green water crashes down on the decks is certainly not experienced so vividly anywhere else as on a sailing vessel which is dependent on its rigging. But this in turn requires of each individual that he be a part of a team; that he show consideration for his environment under exacting conditions, that he bear responsibility for himself and for others and for the ship, and not least he exercise selfdiscipline.*
*The dominating leitmotif which becomes clear for the crew of every sailing ship, "we can only succeed if we all pull together", leads from the direct understanding of the teamwork spirit with its practical necessity to the later realisation of the values of teamwork in theoretical matters. An experience which*

*Arbeit auch in theoretischen Dingen. Eine Erfahrung, die nicht hoch genug eingeschätzt werden kann. Hinzu kommt die Freude am gemeinsamen Tun und Gelingen.*
*Nicht zuletzt aber sei hier durch die vielen internationalen Begegnungen als wertvollste Begleiterscheinung auch des unschätzbaren Beitrages zur Völkerverständigung gedacht. Junge Menschen werden bei dem großen Erlebnis grenzüberschreitender Segelschiffstreffen mit ihren Altersgenossen anderer Nationen zusammenarbeiten und Freundschaften schließen. Und hier sprechen nach kürzester Kontaktaufnahme die Angehörigen unterschiedlichster politischer, gesellschaftlicher oder religiöser Richtungen, einerlei ob aus Ost oder West, Nord oder Süd, stets sehr schnell die gleiche Sprache ohne größere Verständigungsschwierigkeiten!*
*Dem Schiff, das all diese Werte erhalten und fördern soll, wünsche ich für seine Aufgabe stets frischen Wind in die Segel; dem Buch eine weite Verbreitung, um der Bark „Alexander von Humboldt" auch auf diesem Wege zu größtmöglicher Popularität zu verhelfen.*

*Hans Freiherr von Stackelberg*
*Kapitän zur See a. D.*
*und Präsident der STAG.*

*cannot be rated highly enough. This is enhanced by the fun and joy of working together.*
*The most valuable incidental effect which must also be mentioned here is the inestimable contribution towards international understanding achieved by the many encounters with representatives of other countries. In the great experience of the meetings of sailing ships across national frontiers, young people meet their contemporaries from other nations and many lasting friendships are formed. And after the briefest of introductions the members of the most varying political, social or religious tendencies, regardless of whether they come from the north or the south, the east or the west, always very soon speak the same language, without any major communication problems!*
*I wish the ship which is to maintain and strengthen all these values a fresh breeze in her sails at all times, and the book a wide public in order to assist the bark "Alexander von Humboldt" to achieve the greatest possible popularity in this way too.*

*Hans Freiherr von Stackelberg*
*Kapitän zur See a. D.*
*and President of the STAG*

Oben rechts: Das Feuerschiff „Kiel", die jetzige „Alexander von Humboldt", während der siebziger Jahre im Reserveeinsatz auf der Position Fehmarnbelt. Hier hatte es, ebenfalls als „Reserve", sogleich nach der Indienststellung im Dezember 1906 seine Laufbahn als schwimmendes Seezeichen vor der deutschen Küste begonnen.

Top right: The lightship "Kiel", now the "Alexander von Humboldt", on relief duty at the Fehmarnbelt position during the seventies. This is where she started her career as a floating mark off the German coast under the name "Reserve", immediately after her commissioning in December 1906.

Die Bark „Alexander von Humboldt" ist, was Masten, Takelage und Segel, Navigationsausrüstung und Inneneinrichtung betrifft, ein nagelneues Schiff. Der Rumpf aber hat schon ein langes Leben hinter sich. Am Montag, dem 10. September 1906 lief er bei der Werft AG „Weser" in Bremen unter der Baunummer 155 von Stapel.
Das Schiff war das erste einer Serie von vier modernen Feuerschiffen. Anders als seine für die Nordsee bestimmten Schwesterschiffe „Norderney", „Norderney Reserve" und „Amrumbank" war es für die Ostsee vorgesehen. Auftraggeber war die preußische Regierung in Schleswig, abgeliefert wurde es Ende November oder Anfang Dezember 1906 an das zuständige Wasserbauamt Flensburg, und Heimathafen war, wie für alle deutschen Feuerschiffe, die auf der Ostsee westlich Rügens stationiert waren, Sonderburg. Geplant war es von vornherein als Reservefeuerschiff, ein Feuerschiff also, das nicht auf einer festen Station liegt, sondern bestimmte Stammfeuerschiffe während deren jährlicher Werftliegezeit ablöst.
Unter welchem Namen die spätere „Alexander von Humboldt" von Stapel lief, ist nicht ganz geklärt. In Werftunterlagen und

The bark "Alexander von Humboldt" is a brand-new ship as regards masts, rigging and sails, navigational equipment and interior facilities. But the hull has a long history behind her. The vessel was launched on Monday, 10th September 1906, at the AG "Weser" Yard in Bremen as yard number 155.
She was the first of a series of four modern light-vessels. Unlike her sister ships "Norderney", "Norderney Reserve" and "Amrumbank" which were scheduled for the North Sea, this vessel was intended for service in the Baltic Sea. The order for the ships was placed by the Prussian Government in Schleswig, and the vessel was delivered to the competent Wasserbauamt in Flensburg at the end of November or beginning of December 1906. As with all German lightships stationed in the Baltic Sea to the west of Rügen, the port of registry was Sonderburg. The vessel had been planned from the start as a spare lightship, in other words a lightship without an own permanent station but which instead was to relieve certain regular fixed lightships at their proper position during their annual shipyard refit.
Under what name the subsequent "Alexander von Humboldt"

*Presseberichten kommt sie als „Fehmarnbelt" oder „Reserve Fehmarnbelt" vor, was insofern nicht unbedingt falsch ist, als die Position vor Fehmarn die erste war, auf der sie Ende 1906 in Vertretung des Stammfeuerschiffs ausgelegt wurde. Aus diesem Grunde trug sie bereits zum Zeitpunkt der Ablieferung, möglicherweise auch schon beim Stapellauf, auf beiden Seiten des roten Rumpfes die weiße Aufschrift „Fehmarnbelt". Andere Quellen nennen sie „Reserve Sonderburg". In einer amtlichen Liste von 1907 heißt sie nur „Reserve", Sonderburg aber ist als Liegehafen angegeben. Und auf der Schiffsglocke aus dem Jahre 1906, die noch heute auf der „Alexander von Humboldt" gefahren wird, steht „Reserve. Sonderburg", wobei der trennende Punkt das Entscheidende ist. Es ist anzunehmen, daß der Neubau nur „Reserve" hieß, am Heck unter dem Namen aber als Heimathafen Sonderburg stand und das Schiff dann, auch zur Vermeidung von Verwechslungen mit anderen Reserveschiffen, in Behörden- und Besatzungskreisen unter der Bezeichnung „Reserve Sonderburg" geläufig wurde.*

*was launched is not quite clear. In the yard records and press reports she appears as "Fehmarnbelt" or "Reserve Fehmarnbelt", which is not altogether wrong since the position off Fehmarn was her first station at the end of 1906, when she relieved the regular lightvessel there. For this reason she already bore the name "Fehmarnbelt" in white lettering on her red hull at the time of delivery, possibly even at the time of launching too. Other sources call her the "Reserve Sonderburg". In an official list of 1907 she is simply called "Reserve", but Sonderburg was named as home port. And on the ship's bell from the year 1906, which is still carried on the "Alexander von Humboldt" today, "Reserve. Sonderburg" can be read, whereby the full stop separating the two words is of particular importance. It can be assumed that the new ship was simply called "Reserve", but that beneath the name on the stern of the vessel Sonderburg stood as port of registry, and that in order to avoid confusion with other spare light-vessels, the ship became known as the "Reserve Sonderburg".*

Links: Feuerschiff „Kiel" auf seiner Stammposition in der Kieler Bucht, die es bis zur Ablösung durch den neuerbauten Leuchtturm Kiel 1967 innehatte.

Unten: Im Januar 1963 wird das Feuerschiff „Kiel" wegen starken Eisganges von seiner Station zum Kanalhafen Holtenau eingeschleppt. Der Eisbrecher „Preussen" in Hintergrund hat zuvor eine Fahrrinne freigebrochen.

Left: Lightship "Kiel" at her proper position in the Kiel Bay which she held until she was replaced by the newly built Kiel Lighthouse in 1967

Bottom: In January 1963 the lightship "Kiel" was towed from her position to the Canal port of Holtenau due to heavy ice conditions. The ice-breaker "Preussen" in the background had first cut a channel free for her.

Expertengespräche auf dem Zollkreuzer „Bremerhaven" während einer Inspektionsfahrt zu dem auf seiner letzten Position Weser liegenden Feuerschiff im August 1986. Links Kapitän Manfred Hövener, daneben der Danziger Konstrukteur Zygmunt Choreń, der die Umbaupläne zum Segelschiff entwarf. Zweiter von rechts der für die technische Um- und Neuausrüstung des Schiffes verantwortliche Diplomingenieur Prof. Alfred Harms.

Talks between experts on the customs vessel "Bremerhaven" during an inspection trip to the lightship lying at her last position, Weser, in August 1986. On the left Captain Manfred Hövener, next to him the Danzig designer Zygmunt Choreń, who drafted the plans for the conversion to a training sailship. Second from the right is Professor Alfred Harms, Diplomingenieur, who was responsible for the technical conversion and refitting of the ship.

*Mit den Namen von Feuerschiffen ist es ja ohnehin etwas verwirrend, muß man doch unterscheiden zwischen dem Namen des Schiffes, in kleiner Schrift am Heck angebracht, und der Bezeichnung seiner Station, die in großen Lettern an der Bordwand steht. Manchmal sind beide identisch, manchmal nicht. Ganz unübersichtlich wird es, wenn Feuerschiffe beim Wechsel ihrer Position nicht umbenannt werden und dann beispielsweise die „Amrumbank" nicht vor der Amrumbank, sondern auf der Station Deutsche Bucht liegt. Die Reservefeuerschiffe aber änderten bei jedem Positionswechsel die Seitenbezeichnung. Später trugen sie an der Bordwand ständig die Aufschrift „Reserve", der jeweilige Positionsname war auf auswechselbaren Tafeln an den Aufbauten angebracht.*

*The names of lightships are somewhat confusing in any case. One has to distinguish between the name of the vessel applied to the stern in small letters, and the name of her station which is painted in large letters on the ship's sides. Sometimes both names are identical, sometimes they are not. The situation becomes even more involved when lightships are not renamed on changing their station, and for example the "Amrumbank" no longer lies off the Amrum Bank, but instead at the Deutsche Bucht station. However the spare lightships altered the name on their sides with each change of position. Later they always bore the lettering "Reserve" on their sides and the name of the station could be read on interchangeable panels on the upper works.*

Die letzte Feuerschiffsbesatzung nach dem Verlassen der Position während der Fahrt nach Wilhelmshaven. Dritter von links ist ihr letzter Kapitän Horst Welzel. Dabei die Kapitäne Manfred Hövener (mit Mütze) und Wolfgang Vollmer (zweiter von rechts), die zu denen gehören, die später das zum Segler umgebaute Schiff führen werden.

Rechts: Zum letzten Mal im Heimathafen Wilhelmshaven liegt die „Kiel" am Tonnenhof des Wasser- und Schiffahrtsamtes, aus dessen behördlicher Obhut es am 30. September 1986 der STAG zur Nutzung als „Windjammer für die Jugend" übergeben wird.

The last lightship crew after leaving the station on their way back to Wilhelmshaven. Third from the left is the last Captain, Horst Welzel. Next to him the Captains Manfred Hövener (with cap) and Wolfgang Vollmer (second from the right), who will be among those who will later take it in turns to command the vessel when she has been converted into a sailing ship.

Right: In the home port of Wilhelmshaven for the last time, the "Kiel" at the Buoy Yard of the Wasser- und Schiffahrtsamt before being handed over from the official custody of the WSA to the STAG on 30th September 1986 for use as a "windjammer for young people".

WESER RES

Links: Die Flagge der STAG mit dem schwarzrotgoldenen Balkenkreuz weht hoch zwischen den Masten, während das Schiff am 1. Oktober von der Jade zur Weser, vom alten Heimathafen Wilhelmshaven zum künftigen, Bremerhaven, überführt wird.

Oben: Über den Achtersteven verholt das Schiff ins Schwimmdock des Motorenwerks Bremerhaven, des mit dem Umbau beauftragten Werftbetriebes.

Left: The flag of the STAG with the black, red and gold cross flying high between the masts during the transfer of the ship on 1st October from the Jade to the Weser, from her old port of registry Wilhelmshaven to her new home in Bremerhaven.

Top: The vessel being moved sternwards into the floating dock of the Motorenwerk Bremerhaven, the yard commissioned with the conversion.

*Unsere „Reserve" nun war das erste moderne Feuerschiff Deutschlands. Als erstes Schiff verfügte sie über die beiden Konstruktionsmerkmale, mit denen bis heute die Feuerschiffe ausgestattet sind: Sie hatte mittschiffs einen Feuerturm, und sie hatte Maschinenantrieb. Zuvor waren die Feuerschiffe antriebslose Segler, in deren Großmast nachts ein Kranz von drei oder mehr Petroleumlampen vorgeheißt wurde.*
*Die „Reserve" sah allerdings zur Zeit der Indienststellung völlig anders aus als am Ende ihrer Feuerschiffslaufbahn. Abgesehen von der Back hatte sie ein durchlaufendes Deck, der Decksaufbau war schmal und niedrig. Am hinteren Ende des Aufbaus, zwischen Laternen- und Besanmast, befand sich eine*

*Our "Reserve" was Germany's first modern lightship. She was the first vessel to possess the two design features with which lightships are still equipped today: she had a light tower midships, and she was self-propelled. Up to this time the lightships had been sailing ships without any own propulsion and which had hoisted a ring of three or more petroleum lamps up into their main mast at night.*
*However at the time of her commissioning the "Reserve" looked quite different to her appearance at the end of her lightship career. With the exception of the forecastle she had a continuous deck, the deck superstructure being low and narrow. At the rear end of the superstructure, between the*

Unten: Im Dock werden die Bodenschäden besichtigt und repariert, der bis zu dreißig Zentimeter starke Muschelbewuchs wird entfernt.

Rechts: Am Steven des im aufgeschwommenen Dock trockenliegenden Schiffes ist deutlich die mittlere Klüse mit dem Kettenende für den schweren Positionsanker zu erkennen. Die yachtähnliche Unterwasserform mit der V-förmigen Aufkimmung wird das Schiff zu einem guten Segler machen. Die fahrtmindernden Schlingerkiele werden bei einem späteren Werftaufenthalt beseitigt.

Bottom: In the dock the bottom damage is inspected and repaired, and the barnacle crust which is up to thirty centimetres thick is removed.

Right: The middle hawsepipe with the bitter end of the cable for the heavy position anchor can be seen clearly at the bow of the vessel now lying dry in the floated dock. The yacht-like underwater form with the vee-shaped rise of the bottom will make the vessel a good sailing ship. The bilge keels which slow down the rate of travel will be removed during a later period in the yard.

offene Brücke. Zwei Kessel lieferten den Dampf für die von der AG „Weser" gebaute Compoundmaschine. Der senkrechte, etwa ein Meter dicke Schornstein stand vorn auf dem Aufbau zwischen Fock- und Laternenmast, etwa dort, wo sich heute auf dem Manöverdeck der „Alexander von Humboldt" der Magnetkompaß befindet. Nicht nur an Fock- und Besanmast, auch am Feuerturm konnte jeweils ein Gaffelsegel gesetzt werden.
Bei einer Länge zwischen den Loten von 46 m, einer Länge über alles ohne Bugspriet von 53,3 m und einer Breite auf Spanten von 8 m bzw. über alles von 8,04 m war die „Reserve" damals mit 382 BRT und 179 NRT vermessen.
Auf halber Schiffslänge stand ein Feuerturm aus Stahl, wie er seither für Feuerschiffe typisch geworden ist. Zuvor gab es einen solchen Mast nur auf der „Fehmarnbelt" von 1904, diese hatte aber keine Maschine. Der Feuerturm bestand aus einem Mast von 1 m Durchmesser, der eine mit einer Galerie und einem kegelförmigen Dach versehene Leuchtfeuerlaterne von 1,6 m Durchmesser trug. In der Laterne befand sich eine 95 cm

lamp mast and the mizzen mast, there was an open bridge. Two boilers supplied the steam for the compound engine built by the AG "Weser" yard. The vertical funnel, about one metre thick, stood up fore on the superstructure between the foremast and the lamp mast, roughly at the point where the magnetic compass is positioned on the manoeuvre deck of the "Alexander von Humboldt" today. There were facilities for setting a gaff sail not only at the foremast and the mizzen mast, but also at the light tower.
With a length between perpendiculars of 46 m, a length overall without the bowsprit of 53.3 m, a moulded beam of 8 m and an overall beam of 8.04 m, the "Reserve" was rated at the time with 382 G.R.T. and 179 N.R.T. Half way along the ship's length there was a light tower in steel such as has since become characteristic for lightships. Before this time only the "Fehmarnbelt" of 1904 had been equipped with such a tower, but this vessel had no engine. The light tower consisted of a mast with a diameter of 1 m which carried a light lantern with a diameter of 1.6 m provided with a gallery and a conical

Ende Oktober ist das Dock verlassen, die Schäden an der Bordwand sind beseitigt, ebenso die Klüsen für die Feuerschiffsanker. Danach wurden neue Ankerklüsen und Öffnungen für das neu zu installierende Bugstrahlruder geschaffen.

Mit dem Kran wird am 28. Oktober 1986 die Laterne vom Turmmast gehievt, nun ist die „Kiel" endgültig kein Feuerschiff mehr.

At the end of October, the ship has left the dock, the damage to the sidewalls has been repaired and the hawsepipes for the lightship anchors have been removed. New anchor hawsepipes and openings for the installation of the new bow thruster have been made.

Bottom: The lantern was hoisted from the tower mast by a crane on 28th October 1986. The "Kiel" is now definitely no longer a lightship.

Rechts: Das neue Manöverdeck zwischen Mittelaufbau und Poop ist eingezogen und mit Planen abgedeckt, damit darunter die Arbeiten vom Wetter unabhängig weitergehen können.

Unten: Im Frühjahr wird das Schiff wieder eingedockt, die Außenhaut mit Sandstrahlgebläse entrostet und mit grauer Rostschutzfarbe gestrichen.

Right: The new manoeuvre deck between the middle superstructure and the poop has been drawn in and covered with tarpaulins so that work can continue below deck during the winter months regardless of the weather conditions.

Bottom: In spring the ship went back into the dock to have the rust removed from her shell plating by sandblasting and for a coat of grey antirust paint.

*hohe Gürtellinse von 300 mm Brennweite, die zum Ausgleich der Schiffsbewegungen pendelnd aufgehängt war. Das Feuer lag 16 m über Wasser. Damals brannte in diesem Linsenkorb noch keine elektrische Lampe, sondern ein durch Fettgas gespeistes Glühlicht. Um eine schnelle Umrüstung des Feuers auf die unterschiedlichen Kennungen der von dem Reservefeuerschiff wechselnd zu besetzenden Positionen zu ermöglichen, gab es eine entsprechende Zahl von umlaufenden Verdunklungsblenden, deren Drehbewegung wie bei einer Standuhr durch einen Gewichtantrieb in Gang gehalten wurde. Außerdem war das Schiff mit einer preßluftbetätigten Nebelhornanlage, deren Kompressoren mittels einer Dampfpumpe betrieben wurden, und mit einer Einrichtung zur Abgabe von Unterwasserschallsignalen ausgerüstet.*

*So modern auch die Technik des neuen Reservefeuerschiffs war – die Linien seines stählernen Rumpfes folgten bester Segelschiffstradition. Schlank geformt, unter Wasser vorn und achtern scharf geschnitten, mit elegantem Klippersteven versehen und selbst im Hauptspant nicht bauchig wie ein Frachtsegler dieser Zeit, sondern mit V-förmigem Querschnitt gebaut, glich das Feuerschiff in seinen Formen eher einer Yacht. Anders als Windjammer wie „Pamir" und „Passat" sollte ein Feuerschiff ja nicht möglichst viel Ladung aufnehmen, sondern vor seiner Kette liegend auch schwerstes Wetter bestehen können, und dafür eignete sich so ein klipperähnlicher Segelschiffsrumpf eben am besten.*

*roof. The lantern contained a 95 cm high lens with a focal length of 300 mm and was suspended freely so as to compensate the ship's motions. The light was located 16 m above the waterline. In those days the light in this lens cage was not an electric lamp but an incandescent light fuelled with oil gas.*

*In order to allow fast conversion of the light to the differing characteristics of the changing stations taken up by the spare lightship, there was an appropriate number of circulating eclipsing panels which were kept rotating by a weight drive such as is found in a grandfather clock. The ship was also equipped with a foghorn system actuated by compressed air, the compressors being operated by means of a steam pump, and with an installation for emitting underwater sound signals.*

*Modern as the technology of the new lightship was, the lines of her steel hull followed the best sailing ship tradition. Finely moulded, cut sharply fore and aft under water, with an elegant clipper bow, whereby not even the midship section was bulgy like the large cargo sailing vessels of the age but instead had a vee-shaped cross section, the lightship was more like a yacht in her contours. In contrast to windjammers such as the "Pamir" and the "Passat", a lightship does not have to take on as much cargo as possible, but instead must be able to withstand even the heaviest weather lying at anchor, and such a clipper-like sailing vessel hull was quite simply best suited for this.*

*The "Reserve" was a great traveller in her time. She alternated between the stations Kalkgrund at the mouth of the Flensburg*

Nach genauer Berechnung der für das künftige Segelschiff erforderlichen Stabilität werden in mühseliger freiwilliger Arbeit über hundert Tonnen Ballast in Form von Eisenmasseln eingebracht und unten im Schiff fest einbetoniert.

Sämtliches laufendes Gut, die Leinen zur Handhabung der Takelage, werden zwischen Weihnachten und Neujahr von den freiwilligen Helfern im vorübergehend als Reeperbahn dienenden Gang des Hauptdecks maßgerecht zugerichtet und gespleißt.

Rechts: Als erster wird im November 1987 der wie auch die anderen Masten von der Papenburger Sürken-Werft gelieferte Fockmast mit einem Kran auf den zuvor installierten Mastfuß gesetzt. Alter Seefahrtstradition entsprechend hat man auf diesen eine Münze gelegt, die dem Schiff Glück bringen soll. Es ist ein Fünfzigpfennigstück mit dem Bild nach oben, auf dem eine Frau einen jungen Baum pflanzt.

After a precise calculation of the stability required for the future sailing ship, more than a hundred tons of ballast in the form of iron bars were laboriously carried into the ship by volunteers and placed in concrete down at the bottom.

All the running rigging, the lines for handling the rig, are trimmed to size and spliced between Christmas and the New Year by voluntary helpers in the walkway of the main deck which temporarily serves as a rope-maker's alley.

Right: As first mast in November 1987 the foremast, like the other masts supplied by the Sürken-Werft in Papenburg, is placed in position by a crane on the previously installed mast heel. Following an old seafaring tradition a coin intended to bring the ship luck is first placed on the heel, a fifty-pfennig coin with the image of a woman planting a young tree facing upwards.

Fockmast und Bugspriet sind montiert, das Brückenhaus des Feuerschiffes ist verschwunden, das Manöverdeck durchgezogen, nur achtern ragt das halbversenkte neue Kartenhaus darüber hinaus. Auch ist mit dem endgültigen grünen Anstrich begonnen worden, der traditionellen Farbe der berühmten Bremerhavener Rickmers-Segler. Und am Heck steht jetzt der Name „Alexander von Humboldt".

The foremast and bowsprit are mounted, the wheelhouse of the lightship has disappeared, the manoeuvre deck has been drawn through, only the semi-recessed new chart house in the aft part of the vessel rises above it. And work has started on the final green coat of paint, the traditional colour of the famous Bremerhaven Rickmers sailing vessels. The stern already carries the name now, "Alexander von Humboldt".

*Im Laufe der Jahre ist die „Reserve" weit herumgekommen. Abwechselnd lag sie auf den Stationen Kalkgrund in der Einfahrt zur Flensburger Förde und Fehmarnbelt vor der Insel Fehmarn, ab 1912 auch vor Jasmund nordöstlich von Rügen. Im Ersten Weltkrieg wurde die „Reserve" von der Kaiserlichen Marine übernommen und am 28. November 1915 mit einer Marinebesatzung als Kriegsfeuerschiff „Ost" in Dienst gestellt. Zunächst lag sie bei der Gjedsersperre vor der dänischen Küste, ab 1916 nahm sie die Station Libau ein. Als ein deutsches Landungskorps im Oktober 1917 die baltischen Inseln Dagö und Ösel eroberte, legte man mit anderen Feuerschiffen auch die „Reserve" auf den Marschrouten der Landungsverbände aus. Bis November 1918 war sie dann vor Dünamünde in der Rigaischen Bucht stationiert. Sie soll während des Krieges auch mit einem oder zwei Geschützen bestückt gewesen sein.*
*Nach der Volksabstimmung 1920 kam Sonderburg auf der Insel Alsen, der bisherige Liegehafen der „Reserve", an Dänemark, und das Wasserbauamt zog von Flensburg nach Kiel um. Die Feuerschiffe wurden nach Holtenau verlegt, und die „Reserve" erhielt den neuen Namen „Reserve Holtenau".*

*Förde and Fehmarnbelt off the island of Fehmarn, and from 1912 onwards off Jasmund to the northeast of Rügen as well. During the First World War the "Reserve" was taken over by the Imperial Navy and on 28th November 1915 she was commissioned as war lightship "Ost" with a naval crew. She was initially positioned at the Gjedser mine barrage off the Danish coast, and in 1916 she was stationed at Libau. When the Baltic islands of Dagö and Ösel were taken by a German landing force in October 1917, the "Reserve" was among the lightships laid out along the routes taken by the landing troops. Up to November 1918 she was then stationed off Dünamünde in the Bay of Riga. During the war she was allegedly armed with one or two guns.*
*Following the referendum in 1920 Sonderburg on the island of Alsen, which had up to then been the home port of the "Reserve", fell to Denmark and the Wasserbauamt moved from Flensburg to Kiel. The lightships were moved to Holtenau and the "Reserve" was given the new name "Reserve Holtenau". Her tasks remained the same, she continued to guide shipping from changing positions. Between the two World Wars she was stationed at Fehmarnbelt, Kalkgrund (renamed Flensburg in*

Der Kran zieht den Großmast aus der Montagehalle der Werft und setzt ihn hinter den schon stehenden Fockmast auf seinen Fuß. Danach beginnt durch die Taklerfirma Tecklenborg & Kegel die Sicherung mit Stagen und Wanten, zwischen denen dann zumeist in Freizeitarbeit, wie hier von Kapitän Vollmer, die Webeleinen befestigt werden. Nachdem im November auch noch der Besanmast gesetzt ist, stehen alle drei Masten des damit fortan als Bark zu bezeichnenden Schiffes. Gut zu erkennen ist der neu angesetzte Teil des Stevens, der dem Rumpf wieder ein klipperähnliches Aussehen verleiht.

The crane draws the main mast from the yard's assembly shop and sets it on its heel behind the foremast which is already standing. The rigging company Tecklenborg & Kegel then starts to secure the mast with stays and shrouds. Between these the ratlines are attached – generally a spare time occupation as demonstrated here by Captain Vollmer. After installation of the mizzen mast in November too, all three masts are in place and the vessel is now to be called a bark. The newly fitted part of the bow which gives the ship her clipper-like appearance again is shown to advantage here.

Baujahr 1952

In einer der wenigen auf der Welt noch existierenden Segelmachereien, Hinsch & Ruhland in Glückstadt, die noch auf die Ausrüstung großer Rahsegler spezialisiert sind, werden im Winter die fünfundzwanzig Segel mit insgesamt 1036 Quadratmetern für die „Alexander von Humboldt" genäht. Zugleich werden aus seewasserbeständigem Aluminium bei Rönner in Bremerhaven Rahen, Gaffeln und Besanbaum gefertigt. Am 3. März 1988 hängen sie an den Masten, und die Segel werden angeschlagen.

At one of the few sailmakers still existing in the world which specialise in fitting out the large square-riggers, Hinsch & Ruhland in Glückstadt, the twenty-five sails with altogether more than 1036 square metres are sewn for the "Alexander von Humboldt" during the winter. At the same time yards, gaffs and the mizzen boom are made out of seawater-resistant aluminium at Rönner in Bremerhaven. On 3rd March 1988 they are all attached to the masts and the sails are bent to the yards.

Zum ersten Mal verläßt die fertige Bark den Ausrüstungskai des Motorenwerks zur Probefahrt. Im Hintergrund warten Kraftfahrzeuge auf ihre Verschiffung nach Übersee. An der Pier liegt die „Feedermate", eines der modernen Autotransportschiffe der Bremer Reederei Harms. Der Kontrast zwischen einer Vergangenheit, als Segler wie die „Alexander von Humboldt" die Meere beherrschten, und der Seeschiffahrt von heute wird hier augenfällig.

The finished bark leaves the fitting out quay of the Motorenwerk Yard for the first time for sea trials. In the background motor vehicles are waiting for shipment overseas. The "Feedermate", one of the modern automobile transport vessels of the Harms shipping line is moored at the pier. The contrast between a past age, when sailing ships like the "Alexander von Humboldt" ruled the seas, and the shipping of today is conspicuous here.

Über die Toppen geflaggt erscheint die „Alexander von Humboldt" am Tage ihrer Taufe vor der Kulisse der Seestadt Bremerhaven, wo sie jetzt unter ihrem neuen Namen registriert ist.

Dressed with flags, on the day of her christening the "Alexander von Humboldt" appears against the backdrop of the Maritime City of Bremerhaven where she is now registered under her new name.

Bei der Probefahrt zur Abnahme durch den Germanischen Lloyd, der dem TÜV für Kraftfahrzeuge entsprechenden Institution für Schiffe, muß unter anderem ein Drehkreis gefahren werden.

During the trials for certification by Germanischer Lloyd, the inspectorate for shipping, the vessel has to turn a full circle.

*Ihre Aufgaben blieben die gleichen, weiterhin wies sie auf wechselnden Positionen den Schiffen den Weg. Zwischen den beiden Weltkriegen nahm sie die Stationen Fehmarnbelt, Kalkgrund (1925 in Flensburg umbenannt), Adlergrund südwestlich von Bornholm und bis 1925 Jasmund ein. Zeitweise lag sie auch in der Nordsee, unter anderem auf der Station Außeneider.*
*In den Jahren 1933/34 erfuhr die „Reserve Holtenau" einen größeren Umbau. Das Leuchtfeuer und die Nebelhornanlage wurden elektrifiziert. Zur Stromerzeugung baute man zwei Generatoren mit 115 Volt ein, die von zwei Zweizylinder-Zweitakt-Dieselmotoren von Junkers betrieben wurden. Bei Hauptmaschine und Ankerwinde blieb aber der Dampfantrieb erhalten. Ferner erhielt das Schiff ein elektrisches Funkfeuer und Anlagen zur Abstands- und Richtungsbestimmung bei Nebel. 1937 wurden die Besatzungsunterkünfte durch die Einrichtung getrennter Wohn- und Schlafräume verbessert.*
*Im Zweiten Weltkrieg diente die „Reserve Holtenau" offenbar weiter als Reservefeuerschiff oder Wachschiff und war vermutlich mit zwei 2 cm-Flakgeschützen bewaffnet. Anfang 1945 lag sie auf der Station Swinemünde.*

*1925), Adlergrund southwest of Bornholm and up to 1925 at Jasmund. She also took up position in the North Sea from time to time, among other locations at the Ausseneider station.*
*In the years 1933 / 34 the "Reserve Holtenau" underwent a major conversion. The light installation and the foghorn system were electrified. Two generators with 115 volts driven by two two-cylinder two-stroke Junkers diesel engines were installed to generate the power. However steam drive was retained for the main engine and the windlass. The ship was also equipped with an electrical radio beacon and with systems for determining distance and direction in fog. In 1937 the crew's accommodation was improved by the installation of separate living and sleeping quarters.*
*In the Second World War the "Reserve Holtenau" evidently continued to serve as a spare lightship or guard vessel and was presumably armed with two 2 cm anti-aircraft guns. At the beginning of 1945 she lay at the Swinemünde station.*
*The Second World War unexpectedly helped the lightship to obtain a permanent position after her long years of relief service. On 4th May 1945, during the final days of the war, the lightship "Kiel" had been sunk at her post by aircraft*

Arbeit auf den Rahen in den ersten Segeltagen. Jugendliche, Jungen und Mädchen, sind an Bord gegangen und segeln in der Deutschen Bucht das Schiff ein, beraten und unterwiesen von bewährten Fahrensleuten, die einst noch auf großen Frachtseglern wie „Pamir" und „Passat" ihre Seemannslaufbahn begonnen haben.

Work on the yards during the first days under sail. Young people, boys and girls, have come on board and are sailing the ship in the Deutsche Bucht for the first time during the Ascension holidays, with advice and instruction from proven seafarers who started their nautical careers on the large sailing cargo vessels such as the "Pamir" and the "Passat".

*Der Zweite Weltkrieg verhalf dem langjährigen Reserveschiff unverhofft zu einer festen Position. Am 4. Mai 1945 nämlich, in den letzten Kriegstagen, war das Feuerschiff „Kiel" auf seiner Station durch Fliegerbomben versenkt worden. Die „Reserve Holtenau", am 11. September 1946 umbenannt in „Kiel", nahm nun als Stammfeuerschiff diese wichtige Position am Schifffahrtsweg zum Nordostseekanal ein und behielt sie über zwanzig Jahre lang, bis ein fester Leuchtturm ihre Aufgaben übernahm.*

*1950 ersetzte man die alte Dampfmaschine durch einen Dieselmotor, der allerdings weniger Leistung brachte und nur für etwa fünf Knoten Fahrt gut war. Vor dem Feuerturm wurde eine geschlossene Kommandobrücke aufgesetzt. Der Ausbau der Dampfkessel brachte mehr Platz für die Quartiere der Seelotsen, die schon seit 1946 vom Feuerschiff „Kiel" aus auf die einlaufenden Schiffe versetzt wurden und bisher meist in der Messe schlafen mußten. 1954 wurden die alten Funkgeräte gegen eine moderne Funkfeueranlage ausgetauscht.*

*bombs. The "Reserve Holtenau", renamed "Kiel" in the year 1946, now took up this important position as regular lightvessel at the shipping route to the Kiel Canal and retained the station for more than twenty years until a fixed lighthouse took over her duties.*

*In 1950 the old steam engine was replaced by a diesel engine, although this did not develop as much power and was only good for a speed of about five knots. An enclosed bridge was placed in front of the light tower. The removal of the steam boilers left more space for the quarters of the sea pilots, who had transferred to incoming vessels from the lightship "Kiel" since 1946 and who up to then had generally had to sleep in the mess. In 1954 the old radio equipment was replaced by a modern radio beacon system. Lightships were in constant danger of being run into by the ships taking course towards them, and since they lay at anchor they themselves were unable to avoid a collision. The Danish Motor Vessel "Birgitte Skou" had already left a dent in the side of the "Kiel" in*

Am Großtopp ist, bis hinauf zum Skysegel, alles Tuch gesetzt, darunter Royal-, Bram-, Obermars- und Untermarssegel, darüber, zum Fockmast gespannt, Großroyalstagsegel, Großbramstag- und Großstengestagsegel.

On the main mast right up to the skysail all the canvas is set: down below the royal sail, the topgallant, the upper topsail and the lower topsail, up above, set to the foremast, the main royal staysail, the main topgallant staysail and the main topmast staysail.

Links: Die Stimmung an Bord ist gut, wie die Gesichter der auf den Rahen hockenden jungen Leute vermuten lassen.

Left: The atmosphere on board is good, as is reflected by the faces of the young people sitting on the yards.

*Feuerschiffe waren ständig in Gefahr, von den auf sie zusteuernden Schiffen gerammt zu werden, und da sie vor ihrer Kette lagen, konnten sie einer Kollision nicht ausweichen. Bereits im November 1955 hatte das dänische Motorschiff „Birgitte Skou" der „Kiel" die Bordwand eingedrückt, diesmal aber war es noch einmal glimpflich ausgegangen. Doch am 4. Januar 1957 um 08.35 Uhr wurde das Feuerschiff im dichten Nebel von dem Frachter „Satu" aus Helsinki gerammt. Schon um 09.15 Uhr war das Deck achtern nur noch eben aus dem Wasser. Die Kette wurde geslipt, die zwei Lotsenversetzboote versuchten, das Feuerschiff langsam nach Kiel einzuschleppen. Um 10.20 Uhr konnte auch ein inzwischen längsseit gegangener Schlepper mit seinen Pumpen das Wasser nicht mehr halten, das nun schon von den Wohnräumen in den Maschinenraum lief.*

*November 1955, but not too much harm was done that time. However on 4th January 1957 at 08.35 h, the freighter "Satu" from Helsinki ran into the lightship in thick fog. Already, at 09.15 h, the aft deck was barely above water. The chain was slacked away and the two pilots' launches tried to tow the lightship slowly to the port of Kiel. At 10.20 h a tug which had gone alongside in the meantime was unable to hold back the inrushing floods any more with her pumps and water was now running from the living quarters into the engine room. The accompanying vessels then tried to reach shallow water with the sinking "Kiel". At 10.36 h she ran aground in 6 m of water off the Bülk lighthouse. Shortly before noon she suddenly tilted to starboard, snapping both the masts of the tug, and lay in the water with a list of about 45 degrees – but she had been saved from sinking at the very last minute.*

Schon am ersten Segeltag erreicht die „Alexander von Humboldt" bei Windstärke 7 aus Ost zehneinhalb Knoten Fahrt, mehr als sie jemals zuvor unter Maschine gelaufen ist.

Links: Pfingsten in der Nordsee. Sämtliche Segel, mit Ausnahme des Fliegers, sind nun an der vom Konstrukteur vorgesehenen Stelle gesetzt. Später wird man an der Besegelung Änderungen vornehmen und beispielsweise das große Besanbramstagsegel, das das Schiff luvgierig macht, an der Stelle des kleinen Großstengestagsegels führen.

Right on her very first day under sail the "Alexander von Humboldt" made more than ten and a half knots under an easterly wind of force 7, more than she had ever made before under engine power.

Left: Whitsun in the North Sea. Except for the flying jib all the sails are now set in the position planned by the designer. Later the sails will be modified, and for instance the large mizzen topgallant staysail which makes the ship carry the weather helm will be set in place of the main topmast staysail.

Unten: Zwei weibliche Mitglieder der Stammbesatzung beim Umsetzen der Schot des Großstengestagsegels während eines Wendemanövers.

Rechts: Björn aus Flensburg und Kay aus Hamburg beim Holen der Außenklüverschot.

Bottom: Two female trainees shifting the sheet of the main topmast staysail during a tack.

Right: Björn from Flensburg and Kay from Hamburg hauling in the outer jibsheet.

Am Ruder der „Alexander von Humboldt", zweiter von links Kapitän Wolfgang Vollmer.

At the helm of the "Alexander von Humboldt", second from the left Captain Wolfgang Vollmer.

An der Nagelbank beim Aufgeien des Großsegels.

At the pin rack while hauling up the mainsail.

*Man versuchte, mit der sinkenden „Kiel" auf flaches Wasser zu kommen. Um 10.36 Uhr ging sie bei sechs Meter Tiefe vor dem Leuchtfeuer Bülk auf Grund. Kurz vor Mittag kippte sie plötzlich nach Steuerbord, wobei dem Schlepper beide Masten abgeknickt wurden, und lag nun mit etwa 45 Grad Schlagseite im Wasser – das Sinken aber war in letzter Minute verhindert worden.*
*Am 13. Januar 1957 gelang es schließlich, die „Kiel" zu heben. Die Instandsetzung des schwer beschädigten Feuerschiffs wurde auf der Staatswerft in Rendsburg-Saatsee vorgenommen, und man nutzte die zweieinhalbjährige Werftliegezeit zu umfangreichen Modernisierungsmaßnahmen. Das Schiff bekam eine neue Kommandobrücke, die es bis zuletzt behielt. Der Aufbau wurde mittschiffs bis zur Bordwand gezogen und ist in dieser Form noch auf der „Alexander von Humboldt"*

*The authorities finally succeeded in lifting the "Kiel" on 13th January 1957. Repairs to the severely damaged lightship were carried out at the Staatswerft in Rendsburg-Saatsee, and the two and a half year period in the shipyard was also used for extensive modernisations. The ship was given a new bridge which she retained right up to the end. In the midships area the superstructure was drawn right out to the ship's sides and this form has been preserved on the "Alexander von Humboldt". Three modern MWM diesel auxiliary engines for unmanned operation replaced the old engines dating from 1934, and these still supply the ship with electric power today. The conical roof over the lantern with the round knob at its tip disappeared, and in its place a radar scanner thrust its way upwards. In June 1959 the "Kiel" was able to take up her proper position again.*

Die „Alexander von Humboldt" vor Helgoland. Hafen und Reede der Felseninsel in der Deutschen Bucht werden häufig bei Trainingsfahrten angesteuert, die das Schiff ins umliegende Seegebiet unternimmt.

The "Alexander von Humboldt" off Heligoland. The harbour and anchorage of the rocky island in the German Bight are visited frequently by the vessel. From here training voyages are made into the surrounding sea areas.

*erhalten. Drei moderne MWM-Hilfsdiesel für wachfreien Betrieb ersetzten die alten Motoren von 1934. Noch heute versorgen sie das Schiff mit Strom. Auf der Laterne verschwand das kegelförmige Dach mit dem runden Knauf an der Spitze, an seiner Stelle ragte nun eine Radarantenne empor. Am 19. Juni 1959 konnte die „Kiel" wieder ihre Stammposition einnehmen.*
*Auch Eis machte den Feuerschiffen gelegentlich zu schaffen, bestand doch die Gefahr, daß Kette oder Anker dem Eisdruck nicht standhalten und das Feuerschiff vertreibt. So mußte die „Kiel" am 18. Januar 1963 wegen starken Eisgangs mit Schlepper- und Eisbrecherhilfe eingezogen werden und konnte erst am 29. März wieder ausgelegt werden. 1965 brach bei Nordweststurm die Ankerkette, das Schiff vertrieb dreiviertel Seemeilen, bis der Ersatzanker niedergebracht werden konnte. 1966 verlor es bei einer Kollision seinen Bugspriet.*
*Am 5. Juli 1967 verließ die „Kiel" für immer ihre Station: der neue Leuchtturm Kiel, errichtet auf Stahlbetonpontons, die etwa 0,7 Seemeilen von der Feuerschiffsposition entfernt in 13,5 m Wassertiefe auf den Grund abgesenkt worden waren, nahm an diesem Tag seinen Betrieb auf.*

*Ice also occasionally caused trouble for the lightships, since there was a danger of the chain or anchor not standing up to the pressure of the ice so that the lightship would then start to drift. Thus on 18th January 1963 the "Kiel" had to be withdrawn from her station with tug and ice-breaker assistance due to severe floating ice conditions and was unable to return to her proper position until 29th March. In 1965 the anchor chain parted during a northwesterly storm and the ship drifted for three quarters of a mile before the reserve anchor could be dropped. In 1966 she lost her bowsprit as the result of a collision.*
*On 5th July 1967 the "Kiel" left her station for ever: on this day the new Kiel Lighthouse, erected on reinforced concrete pontoons which had been lowered to the sea bed in a water depth of 13.5 m some 0.7 miles from the former lightship position, took up its duties.*
*In the sixties the age of automation in navigation marks set in not only in the Kiel Bay, but also all along the German coast; the age of the manned lightships slowly drew to a close. Lighthouses are substantially cheaper to maintain than lightships and they are indifferent to heavy seas, floating ice and collisions.*

Links: Vorbei an der Kugelbake, dem berühmten Seezeichen, das seit alter Zeit von See her den Beginn des Stromreviers der Elbe kennzeichnet, läuft die „Alexander von Humboldt" nach Cuxhaven ein.

Unten: Auf der Niederelbe vor Brunsbüttel.

Left: Past the Kugelbake, the famous nautical mark which has announced the start of the Elbe estuary since ancient times, the "Alexander von Humboldt" runs into Cuxhaven.

Bottom: On the Lower Elbe, off Brunsbüttel.

*Nicht nur in der Kieler Bucht, an der ganzen deutschen Küste begann in den sechziger Jahren das Zeitalter der Automation im Seezeichenwesen, die Zeit der bemannten Feuerschiffe ging langsam zu Ende. Leuchttürme sind deutlich billiger im Unterhalt als Feuerschiffe, und sie sind unempfindlich gegen See- und Eisgang und gegen Kollisionen.*
*Für das Feuerschiff „Kiel" begann nun ein erneutes Wanderleben als Reserveschiff, seinen Namen aber behielt es. Es löste regelmäßig das Fehmarnbelt-Feuerschiff ab, meist jedoch war es jetzt auf der Nordsee im Einsatz und nahm dort wechselnd die Stationen Elbe 1, Weser, Borkumriff, P 12 (zwischen Helgoland und Wangerooge) und P 8 (nordwestlich der späteren Position Deutsche Bucht) ein. Heimathafen war zunächst weiterhin Kiel, bis das Schiff 1978 ans Wasser- und Schiffahrtsamt Lübeck übergeben wurde.*
*Im Laufe der Jahre wurde die „Kiel" immer wieder modernisiert. Vor den geschwungenen Klippersteven schweißte man einen modernen spitzen Steven, später setzte man ein erhöhtes Poopdeck aufs Achterschiff, und noch später wurde für die Stationskette anstelle der seitlichen eine wuchtige Mittelklüse eingebaut. Auch die Technik wurde ständig erneuert. Das Leuchtfeuer erhielt 1968 einen neuen Drehtisch mit einer Gürtel- und einer Linsenleuchte, die jeweils mit auswechselbaren Blenden für die Kennungen der verschiedenen Feuerschiffspositionen versehen waren. 1970 wurde eine neue kräftige Hauptmaschine mit 510 PS Leistung eingebaut, ein Achtzylinder-Dieselmotor von MAN, der dank guter Pflege und geringer Betriebsstundenzahl noch heute der „Alexander von Humboldt" als Fahrmaschine dient. Auch ein weiterer Hilfsdiesel, den das Schiff 1972 erhielt, ist noch betriebsklar an Bord.*

*For the lightship "Kiel" new years of travel began as she took up service as a spare light-vessel again, but she kept her name. She regularly relieved the Fehmarnbelt lightship, but was now generally on duty in the North Sea, where she alternated between the lightship stations Elbe 1, Weser, Borkumriff, P 12 (between Heligoland and Wangerooge) and P 8 (northwest of the later Deutsche Bucht position). Kiel remained the port of registry until the vessel was transferred to the Wasser- und Schiffahrtsamt Lübeck in the year 1978.*
*In the course of time the "Kiel" underwent a whole series of modernising operations. A modern sharp-edged stem was welded in front of the graceful clipper bow, later an elevated poop deck was placed on the aftship, and later still a massive middle hawsepipe was installed for the station anchor cable in place of the side hawsepipe. The technology on the lightship was constantly updated too. In 1968 the lighting apparatus was equipped with a new turn-table with a dioptric lantern and a lens lantern, each provided with exchangeable panels for the characteristic lights of the differing lightship positions. In 1970 a powerful new main engine with a rating of 510 b.h.p. was installed, an eight-cylinder diesel engine from MAN, which was so well cared for and so little used that it still serves the "Alexander von Humboldt" as propulsion engine today. A further auxiliary which the vessel received in 1972 is still ready for operation on board.*
*After the last lightship, the "Fehmarnbelt", was withdrawn from service in the Baltic Sea, the authorities in Lübeck had no further use for a spare lightship. And so in October 1983 Wilhelmshaven became the new port of registry of the "Kiel", and for the second time she was given an own regular position.*

Neben der Segelpraxis erhalten die Trainees, wie der international geläufige Ausdruck für die Jungen und Mädchen auf Ausbildungsschiffen heißt, Unterricht in Seemannschaft und Navigation – bei gutem Wetter an Deck.

Links: Gemeinsame Abendmahlzeit in der Messe. Auf See essen die Wachen entsprechend ihrer Ablösung jeweils nacheinander.

In addition to sailing practice the trainees – that is the common international expression for the boys and girls on training vessels – are also given instruction in navigation and seamanship, on deck if the weather is good.

Left: Supper together in the mess. At sea the watches take it in turn to eat, depending on when they are relieved.

Zum ersten Mal als Segelschiff verholt die „Alexander von Humboldt" zur Ostsee. Die Fahrt geht bei wechselhaftem Wetter durch den Nordostseekanal, wobei sie, im Regen auf Marschweiden grasende Kühe und bei Sonnenschein das mittelholsteinische Hügelland passierend, eher über Land als durchs Wasser zu gleiten scheint.

The "Alexander von Humboldt" on her way to the Baltic Sea for the first time as a sailing ship. The voyage went through the Kiel Canal under changeable weather conditions, and as she passed cows grazing on the marshy meadows in the rain and the hilly central Holstein landscape in the sunshine, she appeared to be gliding over land rather than through the water.

*Nachdem auf der Ostsee das letzte Feuerschiff, die „Fehmarnbelt", eingezogen war, hatte man in Lübeck keine Verwendung mehr für ein Reservefeuerschiff. So wurde Wilhelmshaven im Oktober 1983 neuer Heimathafen der „Kiel", und zum zweiten Mal bekam sie nun eine Stammposition. Sie löste die außer Dienst gestellte „Amrumbank" auf der Station Deutsche Bucht als Stammfeuerschiff ab. Hier lag sie bis auf kurze Unterbrechungen bis zum 21. Mai 1986, als sie Platz machen mußte für das erste vollautomatische Feuerschiff „FS 1", das keine Besatzung hat und von Land aus fernüberwacht wird. Danach war sie noch im Einsatz auf den Positionen Elbe 1 und Weser. Mittlerweile war aber auch ihre Außerdienststellung beschlossen. Dank der steten, sorgfältigen Pflege, die sie in staatlicher Regie, zuletzt bei der Wasser- und Schiffahrtsdirektion Nordwest Aurich, und durch verantwortungsvolle Feuerschiffbesatzungen genossen hatte, war sie noch vorzüglich erhalten.*

*She relieved the "Amrumbank" which had been put out of commission and took up the post of permanent lightship at the Deutsche Bucht station. Apart from a few brief interruptions she remained here until 21st May 1986, when she had to make way for the first fully automatic lightship "FS 1", which carries no crew on board and is remote-controlled from the shore base. Following this she served terms of duty at the positions Elbe 1 and Weser.*
*In the meantime, however, it had been decided that she too was to be put out of commission. Thanks to the consistent, careful maintenance and attention she had enjoyed under government ownership, most recently with the Wasser- und Schiffahrtsdirektion Nordwest Aurich, and by responsible lightship crews, she was still in excellent condition. It was thus possible to realise the idea of turning the vessel with the perfect windjammer hull completely into a sailing ship, a seagoing bark.*

Nacheinander fährt die „Alexander von Humboldt" unter den Kanalhochbrücken bei Blangenmoor, Hochdonn, Rendsburg und Levensau hindurch, deren lichte Höhe dem Großtopp nur noch wenig Platz läßt.

The "Alexander von Humboldt" passed beneath the high Canal bridges, one after the other, at Blangenmoor, Hochdonn, Rendsburg and Levensau; they do not leave her much space above the main top!

T
404

In einer Regenbö passiert die „Alexander von Humboldt" unweit des Leuchtturms Kiel die Stelle, wo sie früher zwei Jahrzehnte als Feuerschiff auf Station gelegen hat. Wenig später trifft sie die mit vorgeheißtem langem Heimatwimpel von einer Weltreise heimkehrende „Gorch Fock" und begleitet sie mit gekürztem Zeug zur Kieler Förde: die erste Begegnung zwischen dem Segelschulschiff der Bundesmarine und dem einzigen deutschen zivilen Rahsegler.

In a rainy squall close to the Kiel lighthouse the "Alexander von Humboldt" passes the spot where she lay at station as a lightship for two decades. A little later she meets the "Gorch Fock" returning home from a voyage round the world with her long homeward-bound pennant hoisted and accompanies her with shortened sails to the Kiel Förde: the first encounter between the sail training ship of the Bundesmarine and the only German non-naval square-rigger.

*So ließ sich die Idee verwirklichen, aus dem Schiff mit dem perfekten Windjammerrumpf nun ein ganzes Segelschiff zu machen, eine seegehende Bark. Diese zunächst unrealistisch erscheinende Idee war schon Jahre zuvor geboren worden. Ihre beharrlichen Promoter waren vor allem Kapitän Manfred Hövener, Professor an der Hochschule Bremerhaven, und Hans Freiherr von Stackelberg, Kapitän zur See a. D. und früherer Kommandant der „Gorch Fock". Beide gehören zum deutschen Komitee (STAG) der internationalen Sail Training Association. Diese war 1956 in London gegründet worden mit dem Ziel, die damals auf der Welt noch vorhandenen Tiefwassersegler in Fahrt zu erhalten und zu gemeinsamen, der Jugenderziehung und Völkerverständigung dienenden Unternehmungen zu vereinen. Segelschulschiffe verschiedenster Größe aus vielen seefahrenden Nationen haben sich seitdem in den Sommermonaten getroffen, zu Wett- und Geschwaderfahrten, zu Festen in den Häfen. Höhepunkt war jedesmal die populäre „Windjammerparade" aller teilnehmenden Schiffe in den Gasthäfen vor dem jeweiligen Staatsoberhaupt.*

*This idea had been born many years before, but initially appeared unrealistic. Despite this it found persistent advocates in Captain Manfred Hövener, Professor at the Polytechnic in Bremerhaven, and Kapitän zur See a. D. Hans Freiherr von Stackelberg, former Commander of the "Gorch Fock".*
*Both are members of the German Committee (STAG) of the international Sail Training Association. This was founded in London in 1956 with the aim of keeping the blue water sailing ships still remaining in the world afloat and running and of uniting them to joint enterprises serving youth training and international understanding. Since then sail training ships of all sizes from many seafaring nations have met in the summer months for races and cruises in company, for festivals in the ports and harbours. The highlight in each case was the popular "Windjammer Parade" of all the participating vessels in the host ports in front of the Head of State of the host country.*
*The STAG established contact with sponsors who were roused to enthusiasm by the idea of a German windjammer for young*

ALEXANDER VON HUMBOLDT
BREMERHAVEN

Nachdem die Segel mit Geitauen und Gordings von Deck aus aufgegeit sind, steigen die Toppsgäste in die Masten und befestigen die Segel mit Zeisingen an den Rahen. Dabei stehen sie auf den darunter entlanglaufenden starken Leinen, den Fußpferden, und tragen Gurte mit Sicherheitshaken, die bei jeder Arbeit im Rigg an den Jackstagen der Rah eingepickt werden müssen.

After the sails have been hauled up with the clew lines and leech lines from deck, the topmen climb up into the masts and bend the sails to the yards with gaskets. For this they stand on the strong lines running beneath the yards, the foot ropes, and wear harnesses with safety hooks which have to be picked into the jackstays of the yard during all work on the rig.

*Die STAG fand Kontakt zu Förderern, die sich von der Idee eines deutschen Windjammers für die Jugend begeistern ließen. Da war zunächst die Brauerei Beck & Co., die die Jahrhunderte alte Exporttradition der Bremer Brauer erfolgreich fortführt und deren Beck's seit Generationen auf Seeschiffen in aller Welt besonders geschätzt wird. Ebenso ließ sich der erfolgsgewohnte Bremer Unternehmer Egon H. Harms für die Idee begeistern, ist er doch als Reeder und internationaler Autospediteur der Seefahrt eng verbunden. Zusammen mit der STAG gründeten diese beiden Unternehmen die Deutsche Stiftung Sail Training (DSST) und erbrachten das Grundkapital für den Umbau des Feuerschiffs zum Windjammer.*

*people. First of all came the brewery Beck & Co, a company which successfully continues the century-old export tradition of Bremen's brewers and whose Beck's beer has through the generations been in great favour on sea-going vessels all over the world. The long-established Bremen businessman, Egon H. Harms, was also filled with enthusiasm for the idea, especially since he has close connections with ocean shipping as shipowner and international motor vehicle forwarder. Together with the STAG, these two companies established a foundation, the "Deutsche Stiftung Sail Training" (DSST) and put up the capital for the conversion of the lightship into a windjammer.*

Links: Von Flensburg aus, vor deren malerischer Altstadtkulisse an der Schiffbrücke sie hier liegt, unternimmt die „Alexander von Humboldt" bis nach Finnland hinauf Reisen in die Ostsee. Unweit von hier, am Ausgang der Flensburger Förde, in Sonderburg, war einst ihr erster Heimathafen.

Unten: Nachdem sie hier als erstes bundesdeutsches Schiff anläßlich eines Bordempfangs des Bremer Senats zu offiziellem Besuch in der Deutschen Demokratischen Republik gewesen ist, verläßt die „Alexander von Humboldt" den Hafen von Rostock-Warnemünde.

Left: The "Alexander von Humboldt" lying at the Schiffbrücke against the picturesque background of the old part of the City of Flensburg, the starting point for her voyages in the Baltic Sea up as far as Finland. Her first port of registry was not far from here, in Sonderburg at the exit from the Flensburg Förde.

Bottom: After making the first official visit to German Democratic Republic by a vessel from the Federal Republic of Germany and a reception on board given by the Bremen Senate, the "Alexander von Humboldt" leaves the harbour of Rostock-Warnemünde.

Begegnungen mit anderen Schiffen, auf denen ebenfalls wie auf der „Alexander von Humboldt" junge Trainees an Bord sind: mit dem polnischen Dreimaststagsegelschoner „Zawisza Czarny", dessen an Deck angetretene Besatzung mit einem Hurra herübergrüßt, und dem dänischen Vollschiff „Georg Stage".

Encounters with other vessels, manned like the "Alexander von Humboldt" by young trainees: the Polish three-masted staysail schooner "Zawisza Czarny", whose crew has assembled on deck to salute the bark with a loud "Hurrah!", and the Danish full-rigged ship "Georg Stage".

Am 30. September 1986 konnte die STAG das Feuerschiff übernehmen und es für den Umbau in die Deutsche Stiftung Sail Training einbringen. Jedoch zwei Wochen, ehe es seinen ersten Lebensabschnitt als Feuerschiff beschließen sollte, wurde die „Kiel" auf ihrer letzten Position Weser nördlich Wangerooge von dem über 20.000 BRT großen liberianischen Bulkcarrier „Ocean Wind" gerammt. Der Ausguck konnte noch rechtzeitig die Kameraden unter Deck warnen, so kam es nur zu Sachschäden: eingedrückten Platten backbord achtern unter der Wasserlinie und abgeknicktem hinteren Mast. Der Wassereinbruch hielt sich zum Glück in Grenzen, so daß das Schiff mit leichter Schlagseite nach Wilhelmshaven eingeschleppt werden konnte. Nach provisorischer Reparatur beim dortigen Tonnenhof wurde es am 1. Oktober in seinen neuen Heimathafen Bremerhaven überführt. Auf dieser Fahrt trug es den vorläufigen Namen „Confidentia".
Nach den Plänen des berühmten Segelschiffkonstrukteurs der Danziger Werft, Zygmunt Choreń, begann die Umrüstung zur Bark. Ruderhaus, Funkraum und Masten wurden beseitigt, der Feuerturm mitsamt der Laterne demontiert und dem Kieler Schiffahrtsmuseum übergeben. Der Vormast steht jetzt in Wremen auf dem Deich. Die alte schwere Ankerwinde ging ans Schiffahrtsmuseum in Carolinensiel und wurde durch eine für den Segelschiffbetrieb geeignetere ersetzt. Die große Mittelklüse wurde entfernt, die neuen seitlichen Ankerklüsen setzte man höher als die alten. Das Schiff erhielt wieder einen

On 30th September 1986 the STAG was able to take over the lightship and bring it into the "Deutsche Stiftung Sail Training" for the conversion. However two weeks before she was due to finish the first phase of her life as a lightship, at her last position at the Weser station north of Wangerooge, the "Kiel" was rammed by the more than 20,000 G.R.T. large Liberian bulk carrier "Ocean Wind". The look-out was just able to warn his fellow crew members under deck in time, so that only material damage was suffered: plating pressed in beneath the waterline in the aft ship area on the port side and a cracked aft mast. Fortunately the water intake stayed within limits so that it was possible to tow the vessel back to her berth in Wilhelmshaven with a slight list. After temporary repairs there she was transferred to her new home port, Bremerhaven, on 1st October. During this voyage she bore the provisional name "Confidentia".
Work on the conversion to a bark was commenced in accordance with the plans of the famous sailing ship builder from the Danzig shipyard, Zygmunt Choreń. The wheelhouse, radio cabin and masts were removed; the light tower with the lanterns was dismounted and given to the Shipping Museum in Kiel. The foremast now stands on the dyke in Wremen. The heavy, old windlass went to the Shipping Museum in Carolinensiel and was replaced by a model more suitable for the requirements of a sailing vessel. The large, centre hawsepipe was taken out and the new side hawsepipes were set higher

Die „Humboldt" begegnet der schwedischen Gaffelketsch „Atlantica". Im Hintergrund das neue sowjetische Schulschiff „Mir", dessen Pläne ebenso wie die zum Umbau der „Kiel" zum Windjammer „Alexander von Humboldt" von Zygmunt Choreń entworfen wurden.

Links: Die Begegnung mit einem Riesentanker auf See läßt die stattliche „Alexander von Humboldt" geradezu winzig erscheinen und veranschaulicht gleichsam symbolisch die erstaunliche Entwicklung, welche die Seefahrt seit der Segelschiffszeit genommen hat.

The "Humboldt" meets the Swedish gaff ketch "Atlantica". In the background is the new Soviet training ship "Mir", designed by Zygmunt Choreń who also drafted the plans for the conversion of the "Kiel" to the windjammer "Alexander von Humboldt".

Left: The encounter with a giant tanker at sea makes the stately "Alexander von Humboldt" look tiny by comparison and illustrates symbolically the astounding developments undergone by shipping since the age of sailing vessels.

Mit vielen anderen Segelschulschiffen, die sich im schwedischen Marinehafen Karlskrona zu einem internationalen Windjammertreffen versammelt haben, paradiert die „Alexander von Humboldt", die Reihe anführend, vor König Carl XVI. Gustav. Die Besatzung grüßt hinüber zur „Visborg", von deren Deck der König die Parade abnimmt.

Rechts: Besuch in Visby, der traditionsreichen früheren Wikingersiedlung und Hansestadt auf der Insel Gotland. Im Hintergrund der aus dem Mittelalter stammende Dom.

Followed by other sail training ships which have assembled in the Swedish naval port of Karlskrona for an international windjammer meeting, the "Alexander von Humboldt" heads the parade in front of King Carl XVI. Gustav. The crew salutes over to the "Visborg", from whose deck the King reviews the parade.

Right: Visit to Visby, the former Viking settlement and Hanseatic city on the island of Gotland with its wealth of tradition. In the background the cathedral which dates back to the Middle Ages.

*Klippersteven und einen Bugspriet. Freiwillige Helfer schleppten hundert Tonnen Eisenballast an Bord und betonierten ihn in der Bilge ein. Ein neues größeres Ruderblatt und ein Bugstrahlruder verbesserten die Manövriereigenschaften. Zwischen der Poop und dem Mittschiffsaufbau wurde ein durchgehendes Manöverdeck eingezogen, über das sich als höchster gedeckter Aufbau nur das halb versenkte neue Kartenhaus erhebt. Fock-, Groß- und Besanmast wurden aufgesetzt, Rahen angebracht und Segel angeschlagen. Die Masten sind aus Stahl, Rahen, Gaffeln und Baum aus seewasserbeständigem Aluminium. Nur die Skyrah kann gefiert werden, die übrigen Rahen sind fest angebracht. Die Abgasleitungen der Dieselmotoren wurden durch den Besanmast geführt.*

*than the old ones. The ship was given a clipper bow and a bowsprit again. Voluntary helpers dragged mare than a hundred tons of iron ballast on board and laid it in concrete in the bilge. A new, larger rudder blade and a bow thruster improved the manoeuvering capabilities. A continuous manoeuvre deck was drawn between the poop and the midships superstructure, above which only the semi-recessed new chart house rises as highest covered structure. A foremast, main mast and mizzen mast were fitted, yards attached and sails bent. The masts are of steel, the yards, gaffs and boom of seawater-resistant aluminium. Only the skysail yard can be lowered, the remaining yards are attached firmly. The exhaust lines of the diesel engines were passed through the mizzen mast.*

Gäste an Bord. Matrosen vom russischen Schulschiff „Mir", dessen Name im Deutschen „Frieden" bedeutet, sind für einige Tage mit Trainees von der „Alexander" ausgetauscht worden, ein unbürokratischer Akt praktischer Völkerverständigung zwischen jungen Leuten aus Ost und West. Als Gäste waren an Bord auch der Bremer Erste Bürgermeister Wedemeier (der sich während des dortigen Senatsempfangs mit einer Vertreterin der Stadt Rostock unterhält), Bundespräsident von Weizsäcker (im Gespräch mit den Repräsentanten der Deutschen Stiftung Sail Training Egon H. Harms, Leo R. Beckmann und Manfred Hövener), der englische Squarerigger-Kapitän Morin Scott und der schleswig-holsteinische Ministerpräsident Engholm, der sich über das Seemannsgarn amüsiert, das STAG-Präsident Kapitän von Stackelberg zum Besten gibt.

Guests on board. Sailors from the Russian training ship "Mir", whose name means "Peace", have changed places for a few days with trainees from the "Alexander", an unbureaucratic act fostering practical international understanding between young people from East and West. Right: Guests on board included Bremen's 1st Bürgermeister, Mr. Wedemeier (in conversation with a representative of the City of Rostock during the Senate's reception there), the President of the Federal Republic of Germany, Mr. von Weizsäcker (talking to the representatives of the German DSST, Egon H. Harms, Leo R. Beckmann and Manfred Hövener), the English squarerigger captain Morin Scott and the Minister-President of Schleswig-Holstein, Mr. Engholm, who is laughing at the seaman's yarn just being spun by the STAG President Captain von Stackelberg.

Nächste Doppelseite: Die „Alexander von Humboldt" im Hafen von Helsinki zusammen mit (von links) der bulgarischen Barkentine „Kaliakra", dem dänischen Vollschiff „Georg Stage", dem sowjetischem Vollschiff „Mir", dem portugiesischen Viermastgaffelschoner „Creoula", dem deutschen Dreimastgaffelschoner „Großherzogin Elisabeth" und der polnischen Barkentine „Iskra". Mit diesen gehört sie als jüngstes Mitglied zur internationalen Sail Training Association, der Gemeinschaft der sich alljährlich zu Geschwader- und Wettfahrten treffenden Großsegler.

Following double page: The "Alexander von Humboldt" in the harbour of Helsinki together with (from the left) the Bulgarian barkentine "Kaliakra", the Danish full-rigged ship "Georg Stage", the Soviet full-rigged ship "Mir", the Portuguese four-masted gaff-schooner "Creoula", the German three-masted gaff-schooner "Großherzogin Elisabeth" and the Polish barkentine "Iskra". Together with these vessels she belongs to the International Sail Training Association, as youngest member of this community of tall ships which meets annually for races and cruises in company.

11
13
14
15
16

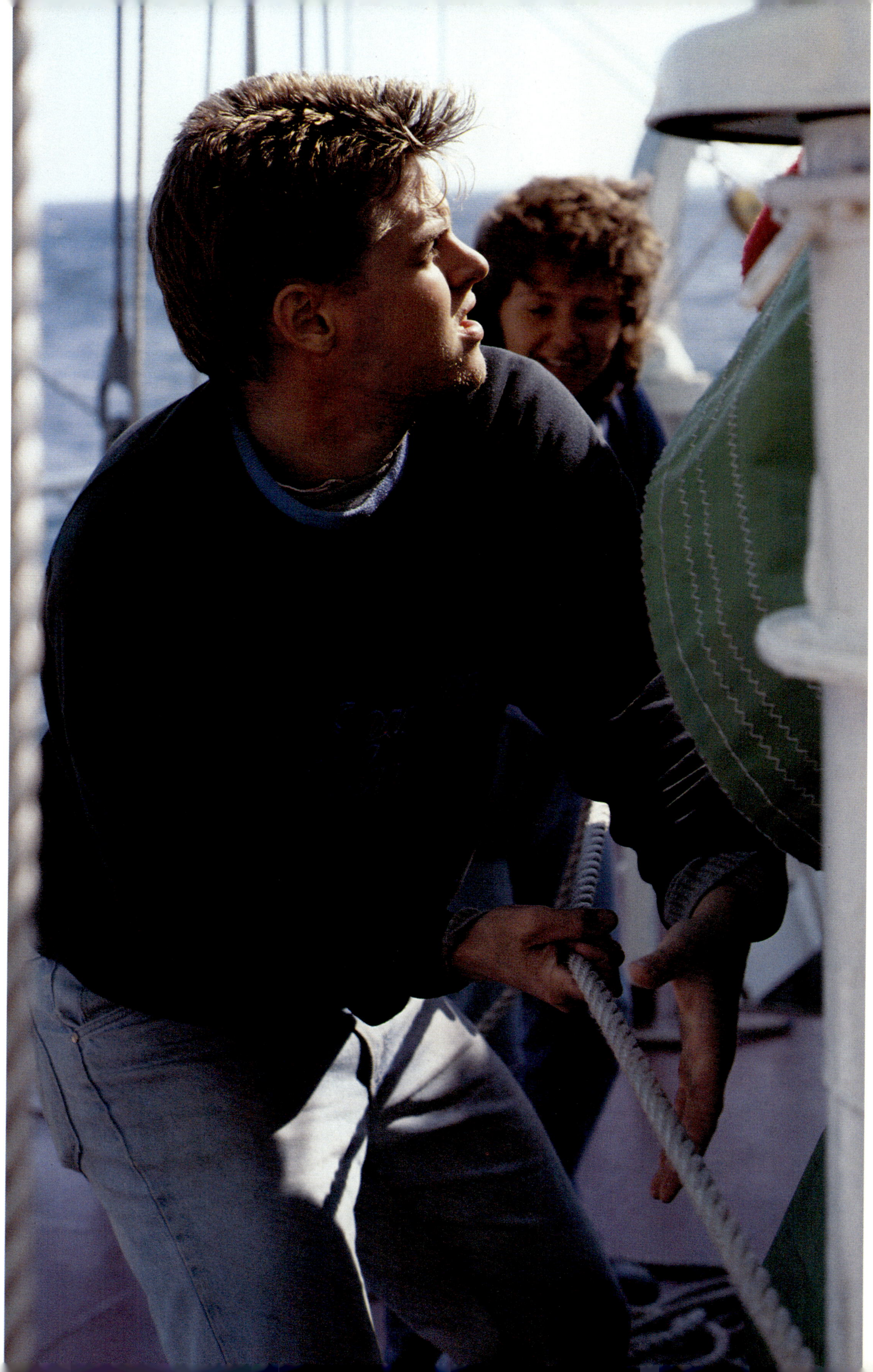

Vorhergehende Doppelseite: Niederholen des Besanstagsegels (links) und des Jagers (rechts).

Segelmanöver:

Oben: Bei einer Wende geht das Schiff mit dem Bug durch den Wind. Der Vortopp steht back, Ober- und Unterbesan sind dichtgeholt, der Großtopp ist bereits rundgebraßt. Wenn das Schiff weiter nach backbord abgefallen ist, werden die Vorsegelschoten über den neuen Bug durchgeholt, und nach dem Kommando „Rund vorn" werden der Vortopp rundgebraßt, der Besan aufgefiert und das Großsegel wieder beigesetzt.

Links: Der Großtopp wird rundgebraßt.

Rechts: Das Fall der Skyrah wird durchgeholt.

Preceding double page: Hauling down the mizzen staysail (left) and the flying jib (right).

Top: During a tack the vessel turns her bow through the wind. The fore-top is aback, the upper and lower spanker are hauled tight, the main top is already braced about. Once the ship has fallen further to port the foresail sheets will be hauled through on the new tack and following the command "Haul taut! Let go and haul!" the fore-top is braced about, the spanker eased away and the mainsail set again.

Left: The main mast is braced about.

Right: The skysail yard halliard is sweated up.

Festmachen der Vorsegel auf dem Bugspriet. Rechts: Blick vom Bugspriet über die Länge des Schiffes nach achtern, Kurs Süd vor Helgoland.

Securing the foresails to the bowsprit. Right: View from the bowsprit aft over the full length of the ship, heading south off Heligoland.

T S G

Vor dem Wachwechsel sind die Mitglieder der neuen Wache an Deck angetreten, nötige Anweisungen werden gegeben und die Vollzähligkeit festgestellt.

Before changing the watch, the members of the new watch assemble on deck, the necessary instructions are issued and full attendance is checked.

Rechts: Kapitän Wittenhagen während einer Manöverbesprechung mit der Stammbesatzung.

Right: Captain Wittenhagen with his nucleus crew discussing tactics prior to a manoeuvre.

*Im Inneren war die Wandlung fast ebenso umfänglich. Hier ging es neben der Umgestaltung der Räume — es mußte Platz für eine sehr viel zahlreichere Besatzung geschaffen werden — vor allem um zusätzliche modernste antriebs-, ruder-, funk- und navigationstechnische Ausrüstung. Sie garantiert dem Schiff einen weit mehr als vorschriftsmäßigen Grad von Sicherheit. Auch der Umweltschutz wurde beachtet: Die neu installierte Abwasserkläranlage erfüllt die Bedingungen von Helsinki-Konvention und MARPOL.*

*Inside the vessel the changes were practically just as radical. In addition to a redesign of the cabins – space for a much larger crew had to be created – particular attention was paid to the installation of the most modern propulsion, rudder, radio and navigational equipment. This guarantees the ship a much higher degree of safety and security than is specified by the regulations. Questions of environmental protection were also considered. The newly installed sewage purification plant conforms with the requirements of the Helsinki Conference and MARPOL.*

Arbeiten an Deck und in der Takelage: Schäkel werden gangbar gemacht, Tauenden gespleißt und Tausendfüßler angebracht, die das Schamfilen, das Scheuern von Segeln an Spieren und Stagen, verhindern sollen.

Work on deck and in the rigging: shackles are worked free, ropes are spliced and bagywrinkles are fitted. These are intended to protect the sails against chafing on spars and stays.

*Diese ganze unübersehbare Arbeit am Werftkai des Motorenwerks in Bremerhaven konnte nur bewältigt werden dank der Hilfe von vielen freiwilligen Mitarbeitern, Kapitänen, Seelotsen, Ingenieuren, Schiffbauern, Handwerkern, pensionierten Seeleuten, Hobbyseglern und vielen begeisterten jungen Leuten, die unzählige unbezahlte Arbeitsstunden auf dem Schiff geleistet haben. Ferner vermittelte das Berufsbildungs- und Technologiezentrum der Kreishandwerkerschaft Bremerhaven mit Unterstützung des Arbeitsamtes Auszubildende zur praktischen Arbeit auf das Schiff. Überdies stellte das Arbeitsamt im Rahmen von Arbeitsbeschaffungsmaßnahmen weitere Hilfe bereit. Zahlreiche Unternehmen erbrachten beträchtliche Geld- und Sachleistungen.*

*All this incalculable volume of work at the quay of the Motorenwerk in Bremerhaven could only be managed thanks to the aid of large numbers of voluntary staff, captains, sea pilots, engineers, shipbuilders, craftsmen, retired seamen, hobby sailors and many enthusiastic young people who invested countless unpaid manhours. In addition the Vocational Training Centre of the Bremerhaven Regional Association of Trade Handicrafts and the Labour Exchange seconded apprentices to the project. Many firms supplied substantial donations in cash and in kind.*

TS
G 404

Vorhergehende Doppelseite: Der Wind ist stärker geworden, bis auf Vorsegel, Untermarssegel, zwei Stagsegel und den Besan ist alles Tuch geborgen.

Links: Hoch spritzt die Gischt auf, wenn im Sturm der Bug tief in jede anrollende See einsetzt.

Oben: Es hat aufgebrist, mit Mühe werden auf dem Bugspriet die Vorsegel geborgen.

Previous double page: The wind has become stronger, apart from head sails, the lower topsails, two staysails and the spanker, all the canvas has been hauled in.

Left: The spray splashes up high when the bow delves into the waves in stormy weather.

Top: The wind has freshened, on the bowsprit it is not so easy to take in the foresails.

Am 3. März 1988 schließlich war es soweit. Zur Abnahme durch den Germanischen Lloyd (einer Art TÜV für Schiffe) verließ die jetzige Bark zum ersten Mal Ausrüstungskai und Schleuse. Draußen auf der Außenweser fanden die Erprobungen statt, insbesondere auch die von Maschine und Ruderanlagen. Alles verlief zufriedenstellend.
Gesegelt wurde noch nicht. Das geschah beim „Eintrainieren" der ersten Stammannschaft aus alten Fahrensleuten der Pamir-Passat-Vereinigung und segelbegeisterten Jugendlichen zu Himmelfahrt 1988 im Seegebiet um Helgoland. Überraschend gut, das bestätigten auch die Besichtiger des Germanischen Lloyd, erwiesen sich die Segeleigenschaften. 10,5 Knoten lief die „Alexander von Humboldt" bei Windstärke 7 aus Ost bei strahlendem Sonnenschein, schneller als jemals zuvor mit ihrer Maschine.

Finally, on 3rd March 1988, the day arrived. The vessel, now a bark, left the fitting-out quay and lock for the first time for acceptance tests by Germanischer Lloyd. The trials, especially of the engine and steering installations, were conducted at sea in the Outer Weser. Everything ran satisfactorily.
However the ship was not yet sailed. The sails were set for the first time during the "initial training" of the first nucleus crew made up of experienced seafarers from the Pamir-Passat Association and young sailing enthusiasts in the sea area around Heligoland during the Ascension holiday in 1988. The sailing qualities were surprisingly good, this was also confirmed by the surveyors from Germanischer Lloyd. The "Alexander von Humboldt" ran 10.5 knots under an easterly wind, force 7, on a bright sunny day – faster than she had ever run before under engine power.

Bei rauhem Wetter wird jede Hand gebraucht.

Nächste Doppelseite: Lars aus Brunsbüttel und Jochen aus München holen den Großsegelhals.

Every hand is needed in heavy weather.

Next double page: Lars from Brunsbüttel and Jochen from Munich hauling the mainsail tack.

Vorhergehende Doppelseite: Blick vom Vortopp aufs Deck und nach achtern auf die Rahsegel des Großmastes und das davor geheißte Großstengestagsegel (links), sowie vom Bugspriet auf die Back, wo ständig ein Ausguck Wache geht (rechts).

Unten: Bei völliger Windstille spiegelt sich der Bug im Wasser, der fallende Anker verwischt das Bild.

Rechts: Im Sommer auf See. Das Schiff hat beigedreht, Gelegenheit zum Schwimmen von Bord aus.

Previous double page: View from the fore-top onto the deck and aft to the square sails of the main mast and the main topmast staysail which is hoisted in front of it (left), and from the bowsprit to the forecastle where a lookout is constantly on watch (right).

Bottom: In a complete calm the bow is mirrored in the water, but the picture vanishes when the anchor drops.

Right: At sea in the summer. The ship has hove to, a chance to have a swim from the deck.

Decksarbeit und Segelmanöver in frischer Seeluft sorgen für gesunden Schlaf, dem man sich bei sonnigem Wetter am liebsten an Deck hingibt.

Work on deck and sailing manoeuvres in the fresh sea air guarantee a sound sleep, which is enjoyed best on deck in sunny weather.

Unten: Monika aus Hamburg und Andrea aus Berlin auf dem Bugspriet. Die auf jedem Segel angebrachte genaue Beschriftung ist deutlich zu erkennen, am Fockstag der braune Tausendfüßler, der die Fock vor dem Schamfilen schützen soll.

Rechts: Bei frischer Backstagsbrise in der Nordsee. Im Sonnenlicht glänzen die aus besonders dauerhaftem Kunstfasertuch genähten grünen Segel.

Bottom: Monika from Hamburg and Andrea from Berlin on the bowsprit. The exact lettering applied to each sail can be recognised clearly here; on the forestay the brown bagywrinkle which protects the foresail from chafing.

Right: Under a fresh quartering breeze in the North Sea. The green sails sewn from a particularly durable synthetic material gleam in the sunlight.

G - 4 0 4
04

*Am Freitag vor Pfingsten, zum Auftakt der Nordseewoche 1988, wurde das Schiff in Bremerhaven getauft auf „Alexander von Humboldt" nach dem berühmten deutschen Naturforscher und Weltreisenden, dem Mitbegründer der Berliner Universität, dessen Name wie kaum ein anderer geeignet ist, sein Heimatland auf künftigen Reisen in der Welt zu vertreten.*
*Konstanze-Caroline Freifrau von Humboldt, eine Nachfahrin von Alexanders nicht minder bedeutendem Bruder Wilhelm, vollzog im Beisein von Kindern aus dem Kreise der Stifter den Taufakt. Danach ging das Schiff unter Führung seines ersten Kapitäns Rudolf Wittenhagen wieder hinaus in die Nordsee — zu Beginn eines neues Lebensabschnittes: als Windjammer für die Jugend.*
*Und im gleichen Sommer, in dem die „Alexander von Humboldt" die ersten Reisen unter Segeln machte, fand die Epoche der bemannten Feuerschiffe an der deutschen Küste endgültig ihren Abschluß. Das letzte von ihnen verließ im Juli 1988 mit seiner Besatzung die Position Borkumriff und wurde durch eine vollautomatische Nachfolgerin ersetzt.*

*On the Friday before Whitsun, at the start of the Nordseewoche 1988, the ship was christened in Bremerhaven with the name "Alexander von Humboldt" after the famous German naturalist and traveller, the co-founder of the University of Berlin, whose name is better suited than hardly any other to represent his native country on future voyages throughout the world.*
*Baroness Konstanze-Caroline von Humboldt, a descendant of Alexander's brother Wilhelm who is of no less historical importance, performed the christening, supported by children of the DSST members. And then the ship sailed out into the North Sea again, under her first captain, Rudolf Wittenhagen, to start a new phase of life as a windjammer for young people.*
*And in the same summer in which the "Alexander von Humboldt" made her first voyages as a sail training ship, the age of the manned light-vessels on the German coast finally drew to a close. The last of them with her crew left the Borkumriff position in July 1988 and was replaced by a fully automatic successor.*

Gute Laune herrscht vor auf der „Alexander von Humboldt". Ein Grund dafür mag die ungewöhnliche sympathische Gemeinschaft sein, die hier zwischen segelbegeisterten jungen Leuten und „alten Seebären" entsteht. Diese sind größtenteils Mitglieder der Pamir-Passat-Vereinigung, in der sich Seeleute zusammengefunden haben, die in ihrer Jugend noch aktiv auf großen Handelsseglern gefahren sind. Zu ihnen gehören auch die ehemaligen „Passat"-Fahrer Hein Behn, jetzt Sailing Master auf der „Alexander von Humboldt", Kapitän Rudolf Wittenhagen und Bootsmann Egon Thomsen (von links nach rechts), die sich an Bord ebenso wohl fühlen wie die jungen Trainees (oben): Stefanie aus Bremerhaven und Kathrin aus Hamm, Marc aus Köln, Jan-Peter aus Hamburg, Torsten aus Oldenburg und Anna aus Danzig.

A good mood prevails on the "Alexander von Humboldt". One reason for this might be the exceptionally congenial community spirit which develops between the enthusiastic young people and the old "sea bears". Most of the latter are members of the Pamir-Passat Association which is made up of seamen who spent part of their youth in active service on the large sailing freighters. They also include the former "Passat" crewmen Hein Behn, now Sailing Master on the "Alexander von Humboldt", Captain Rudolf Wittenhagen and the bosun Egon Thomsen (from left to right), who feel just as at home on board as do the young trainees (above): Stefanie from Bremerhaven and Kathrin from Hamm, Marc from Cologne, Jan-Peter from Hamburg and Torsten from Oldenburg, Anna from Danzig.

TSG - 404

Vorhergehende Doppelseite: Zwischen den flachen Granitfelsen der Schären nähert sich die „Alexander von Humboldt“ der Einfahrt von Mariehamn auf den finnischen Ålandinseln. Rechts im Hintergrund die alte Lotsenstation Kobba Klintar, links auf der Felseninsel das Erinnerungsmal für den Laeisz-Segler „Plus“, der hier in einem Schneesturm gesunken ist.

Oben: Die Toppsgäste sind auf die Rahen geklettert und befestigen die Segel. Wie die alten Klipper ist die „Alexander von Humboldt“, nach und nach das Tuch wegnehmend und somit die Fahrt mindernd, ohne Maschine bis vor die Stadt gesegelt und zu Anker gegangen.

Rechts: Ihren Liegeplatz in Mariehamn bekommt die „Alexander von Humboldt“ neben der Viermastbark „Pommern“, die hier nach beendeter aktiver Laufbahn alljährlich das Interesse von vielen tausend Besuchern findet. Außer der in New York als Museumsschiff liegenden „Peking“, der in Travemünde zu besichtigenden „Passat“ und der als russisches Schulschiff noch in Fahrt befindlichen „Krusenstern“, der ehemaligen „Padua“, gehört sie zu den letzten noch existierenden „Flying P-Linern“, den berühmten großen Frachtseglern der Hamburger Reederei Laeisz. Auch die „Pamir“ war eine von ihnen, deren tragischer Verlust im Jahre 1956 praktisch das Ende der Handelsschiffahrt unter Segeln bedeutete und das Verschwinden der stolzen Windjammer von den Meeren. Kein großes Segelschiff fährt mehr Fracht, aber eine Flotte von Segelschulschiffen, bemannt mit jungen Leuten, führt in vielen seefahrenden Ländern ihr Erbe fort, unter ihnen das einstige Feuerschiff, der Windjammer für die Jugend „Alexander von Humboldt“.

Preceding double page: Between the flat granite rocks of the skerries the "Alexander von Humboldt" approaches the entrance to Mariehamn on the Finnish Åland islands. On the right in the background the old pilot station Kobba Klintar, to the left on the rock island the memorial for the Laeisz sailing vessel "Plus" which sank in a snow-storm here.

Top: The topmen have climbed up to the yards and are bending the sails. Like the old clippers the "Alexander von Humboldt" has sailed right up to the town without using engine power, gradually hauling down canvas to reduce speed, and has anchored.

Right: She is then given a berth in Mariehamn next to the four-masted bark "Pommern", which here excites the interest of many thousands of visitors every year now that she has completed her active life. Along with "Peking" lying in New York as a museum ship, the "Passat" which can now be viewed in Travemünde, and the "Krusenstern" – formerly the "Padua" — which still sails as the Russian sail training vessel, she belongs to the last "Flying-P-Liners" still in existence, the famous large sailing cargo vessels of the Hamburg Laeisz line. The "Pamir" was one of these too, and her tragic loss in the year 1956 virtually meant the end of merchant shipping under sail and the disappearance of the proud windjammers from the seas. None of the large sailing ships carry cargo anymore, but a fleet of sail training vessels, crewed by young people, continues their heritage in many seafaring countries, though as "windjammers for young people" under quite different conditions, and one of these is the former lightship "Alexander von Humboldt".

POMMERN
ALEXANDER VON HUMBOLDT
BREMERHAVEN

# Generalplan
# General Plan

## *Segelriß*
## *Sail Plan*

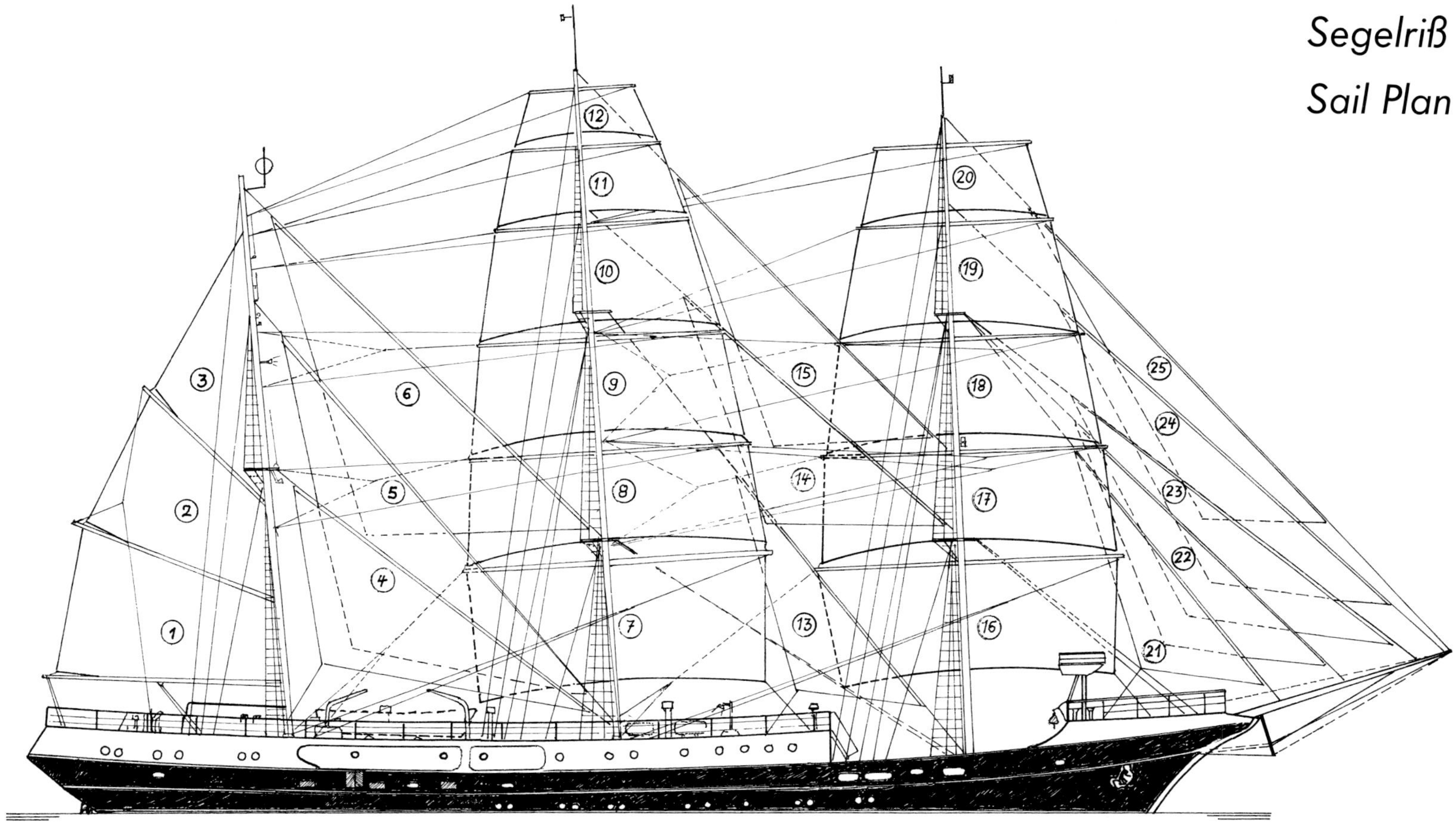


### *Generalplan / General Plan*

1 Navigationsraum / Chart room
2 Kabelgatt / Rope store
3 Messe / Mess
4 Dutymesse / Duty mess
5 Proviant / Provision store
6 Waschraum, Toilette / Washroom, lavatory
7 Kombüse / Galley
8 Messe / Mess
9 Pantry / Pantry
10 Waschraum, Waschmaschine / Washroom, washing machine
11 Gang / Passageway
12 Toilette / Lavatory
13 Akkuraum / Storage batteries
14 Toilette / Lavatory
15 Toilette / Lavatory
16 Abgasschacht / Gas trunk
17 Store / Store
18 Store / Store
19 Ölzeug / Oilskins
20 Waschraum, Toilette / Washroom, lavatory
21 Kapitänssalon / Captain's saloon
22 Kapitänskajüte / Captain's cabin
23 Kapitänssalon / Captain's saloon
24 Unteres Kabelgatt / Lower rope store
25 Kettenkasten / Chain locker
26 Umkleideraum / Changing room
27 Waschraum / Washroom
28–30 Kabinen / Cabins
31 Toilette / Lavatory
32–37 Kabinen / Cabins
38 Waschraum, Toilette / Washroom, lavatory
39–42 Kabinen / Cabins
43 Store / Store
44 Schalttafel / Switchboard
45 Dieselaggregate / Diesel-generator sets
46 Dieselaggregat / Diesel-generator set
47 Store / Store
48 Trockenraum / Drying room
49 Hilfsdiesel / Auxiliary diesel engine
50 Last 3 / Hold 3
51 Last 4 / Hold 4
52 Bugstrahlruderraum / Bow thruster room
53 Last 1 / Hold 1
54 Last 2 / Hold 2
55 Store / Store
56 Hilfsmaschinenraum / Auxiliary engine room
57 Maschinenraum / Engine room
58 Last 3 / Hold 3

### *Segelriß / Sail Plan*

| | | |
|---|---|---|
| 1 Unterbesan | lower spanker | 43,8 qm / sq m |
| 2 Oberbesan | upper spanker | 33,7 qm / sq m |
| 3 Besantoppsegel | gaff topsail | 27,6 qm / sq m |
| 4 Besanstagsegel | mizzen staysail | 40,7 qm / sq m |
| 5 Besanstengestagsegel | mizzen topmast staysail | 63,8 qm / sq m |
| 6 Besanbramstagsegel | mizzen topgallant staysail | 63,2 qm / sq m |
| 7 Großsegel | mainsail, main course | 74,8 qm / sq m |
| 8 Großuntermarssegel | lower main topsail | 55,6 qm / sq m |
| 9 Großobermarssegel | upper main topsail | 56,6 qm / sq m |
| 10 Großbramsegel | main topgallant sail | 36,7 qm / sq m |
| 11 Großroyalsegel | main royal | 20,8 qm / sq m |
| 12 Großskysegel | main skysail | 11,2 qm / sq m |
| 13 Großstengestagsegel | main topmast staysail | 22,6 qm / sq m |
| 14 Großbramstagsegel | main topgallant staysail | 35,7 qm / sq m |
| 15 Großroyalstagsegel | main royal staysail | 41,9 qm / sq m |
| 16 Fock | foresail, fore course | 67,6 qm / sq m |
| 17 Voruntermarssegel | lower fore topsail | 55,6 qm / sq m |
| 18 Vorobermarssegel | upper fore topsail | 56,6 qm / sq m |
| 19 Vorbramsegel | fore topgallant sail | 36,7 qm / sq m |
| 20 Vorroyalsegel | fore royal | 20,8 qm / sq m |
| 21 Vorstengestagsegel | fore topmast staysail | 23,6 qm / sq m |
| 22 Innenklüver | inner jib | 30,1 qm / sq m |
| 23 Außenklüver | outer jib | 40,1 qm / sq m |
| 24 Jager | flying jib | 41,2 qm / sq m |
| 25 Flieger | jib-o-jib | 34,6 qm / sq m |

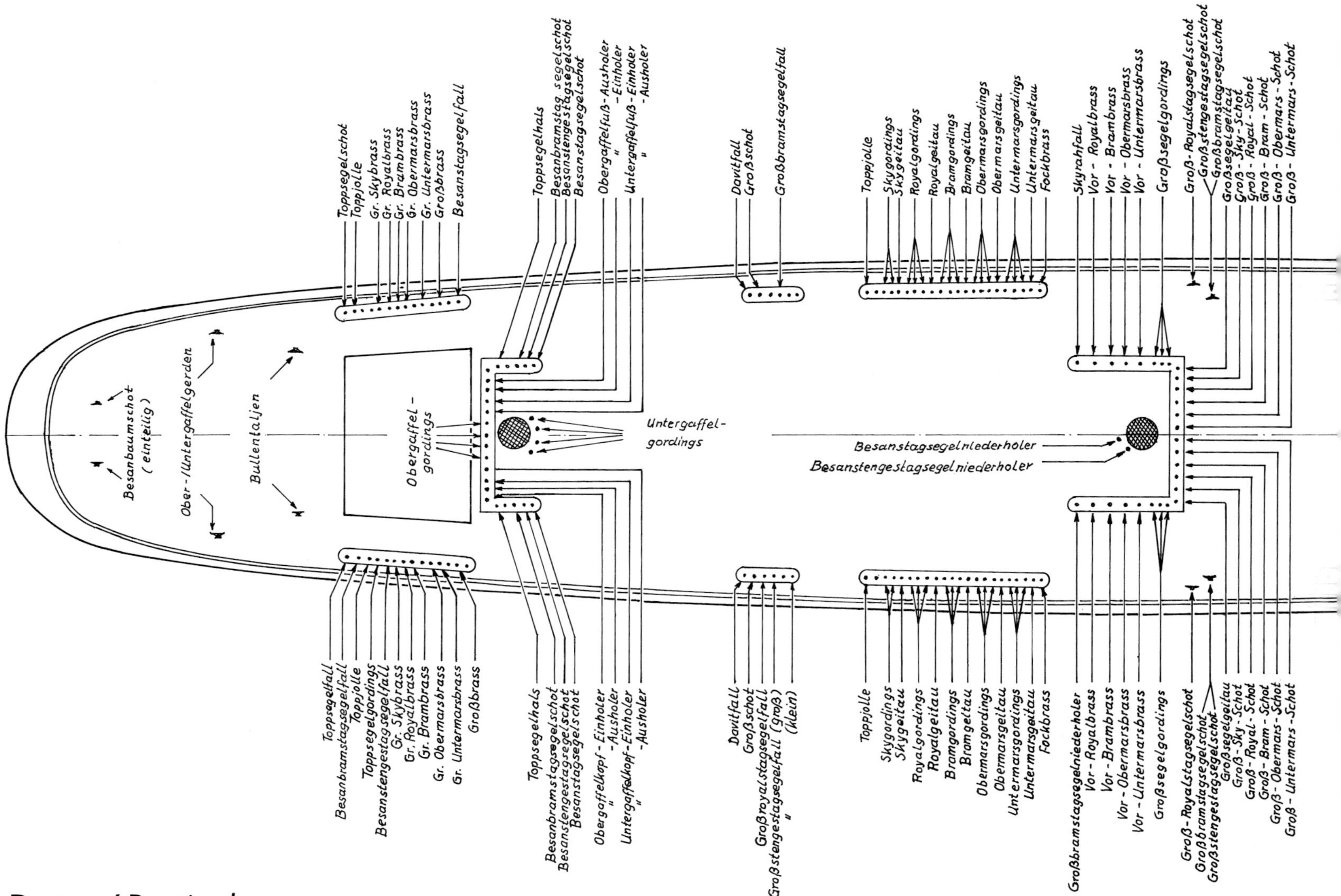


# Daten / Particulars

"Alexander von Humboldt" ex "Confidentia" ex "Kiel" ex "Reserve Holtenau" ex "Reserve" oder/or "Reserve Sonderburg"

| | |
|---|---|
| Heimathafen / Port of registry | Bremerhaven |
| Segelzeichen / Sail number | TS G 404 |
| Rufzeichen / Call signal | DFAW |
| Stapellauf / Date launched | 10. September 1906 |
| Ablieferung / Date delivered | November / Dezember 1906 |
| Bauwerft / Building yard | AG "Weser", Bremen |
| Baunummer / Yard number | 155 |
| Umbau zur Bark / Converted to a bark | Oktober 1986 bis Mai 1988 |
| Umbauwerft / Conversion yard | Motorenwerk Bremerhaven |
| Umbaunummer / Conversion number | 97 210 |
| Länge über alles mit Bugspriet / Length overall incl. bowsprit | 62,55 m |
| Länge über alles ohne Bugspriet / Length overall without bowsprit | 54,00 m |
| Länge zwischen den Loten / Length between perpendiculars | 46,60 m |
| Breite auf Spanten / Moulded beam | 8,00 m |
| Breite über alles / Overall beam | 8,02 m |
| Seitenhöhe bis Hauptdeck / Moulded depth up to the main deck | 5,52 m |
| Seitenhöhe bis Manöverdeck / Moulded depth up to the manoeuvre deck | 7,72 m |
| Kielhöhe / Depth of keel | 0,28 m |
| Tiefgang leer / Light draught | 4,30 m |
| Tiefgang vollbeladen / Draught fully laden | 4,88 m |
| Gewicht leer / Empty weight | 660 t |
| Gewicht vollbeladen / Weight fully laden | 829 t |
| Raumgehalt / Tonnage | 396 BRZ / GT |
| Mastenhöhe über Wasserlinie für 4,80 m Tiefgang / Mast height above water level at a draught of 4.80 m: | |
| Fockmast / Foremast | 29,90 m |
| Flaggenstock / Flag-staff | + 2,00 m |
| Großmast / Main mast | 32,00 m |
| Flaggenstock / Flag-staff | + 2,00 m |
| Besanmast / Mizzen mast | 28,10 m |
| Segelfläche / Sail area | 1035,6 qm / sq m |
| Hauptmotor / Main engine | 8 Zylinder 4-Takt MAN / 8-cylinder 4-stroke MAN |
| Leistung / Rating | 375 kW / 510 PSe / h.h.p. |
| Umdrehungen / Revolutions per minute | $n = 1600\ min^{-1}$ |
| Schiffsklasse / Class of ship: | ✢ 100 A 4 E Sailing Yacht (Schiffskörper / hull)<br>✢ MC (Maschine / engine)<br>Safety Assessment des Germanischen Lloyd |
| Stammbesatzung / Nucleus crew | 15 |
| Trainees | 45 |

Generalplan: Original-Umbauplan von Zygmunt Choreń. Segelriß: nach dem Plan von Zygmunt Choreń gezeichnet von Peter Knuth. Belegplan: nach Angaben von Kpt. Rudolf Wittenhagen gezeichnet von Peter Knuth. Linienriß: nach Plänen des Wasserstraßen- und Maschinenamts Rendsburg aus dem Jahre 1950 gezeichnet von Peter Knuth. Daten: basierend auf den Umbauplänen, gültig zur Zeit der Indienststellung.

General Plan: original conversion plan by Zygmunt Choreń. Sail Plan: drawn by Peter Knuth from the plan drafted by Zygmunt Choreń. Belaying Plan: drawn by Peter Knuth in accordance with the directions of Capt. Rudolf Wittenhagen. Lines Drawing: drawn by Peter Knuth from plans of the Wasserstraßen- und Maschinenamt Rendsburg dating from the year 1950. Particulars: based on the conversion plans, valid at the time of commissioning.

## Belegplan

## Belaying Plan

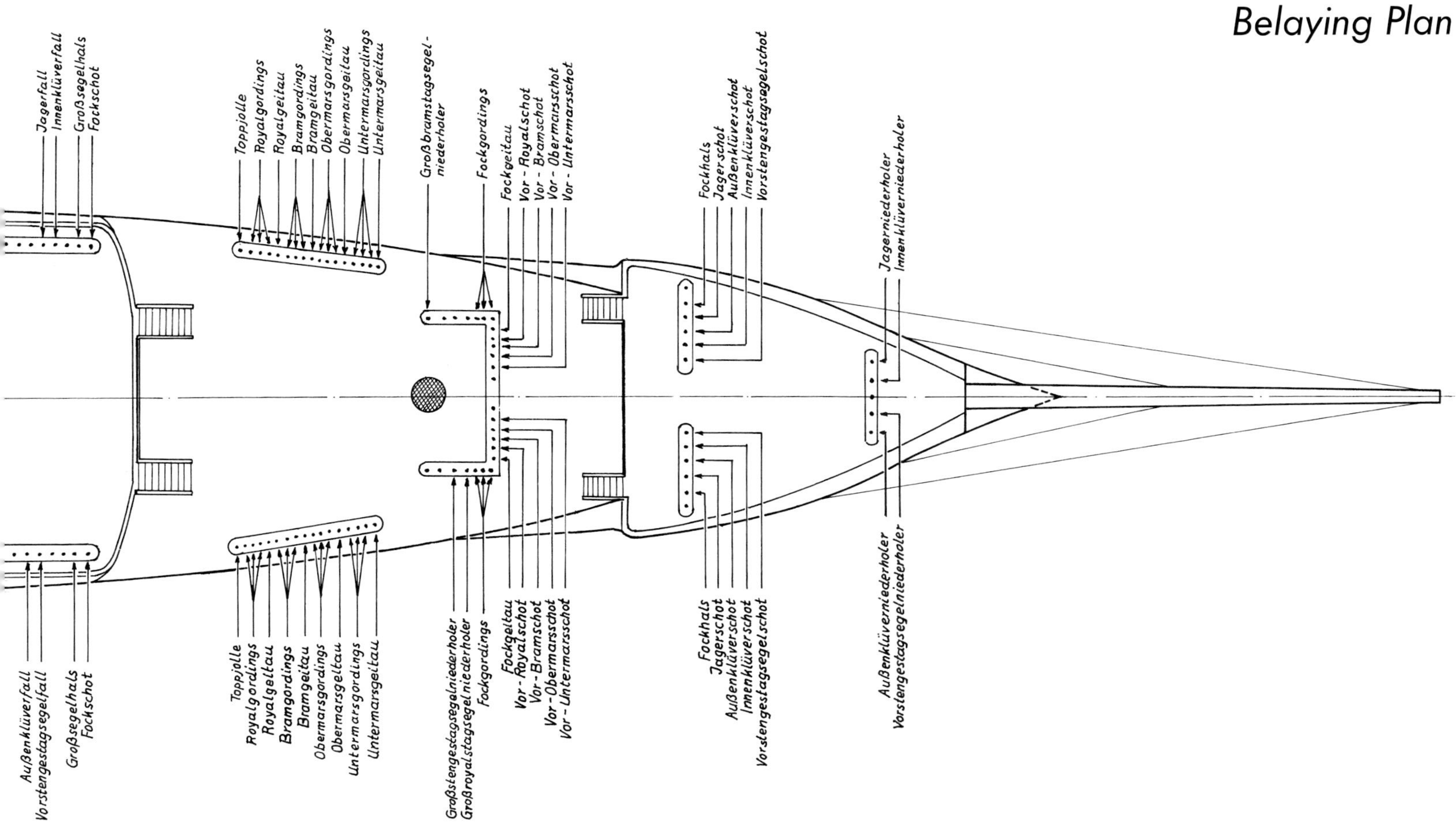


## Linienriß

## Lines Drawing

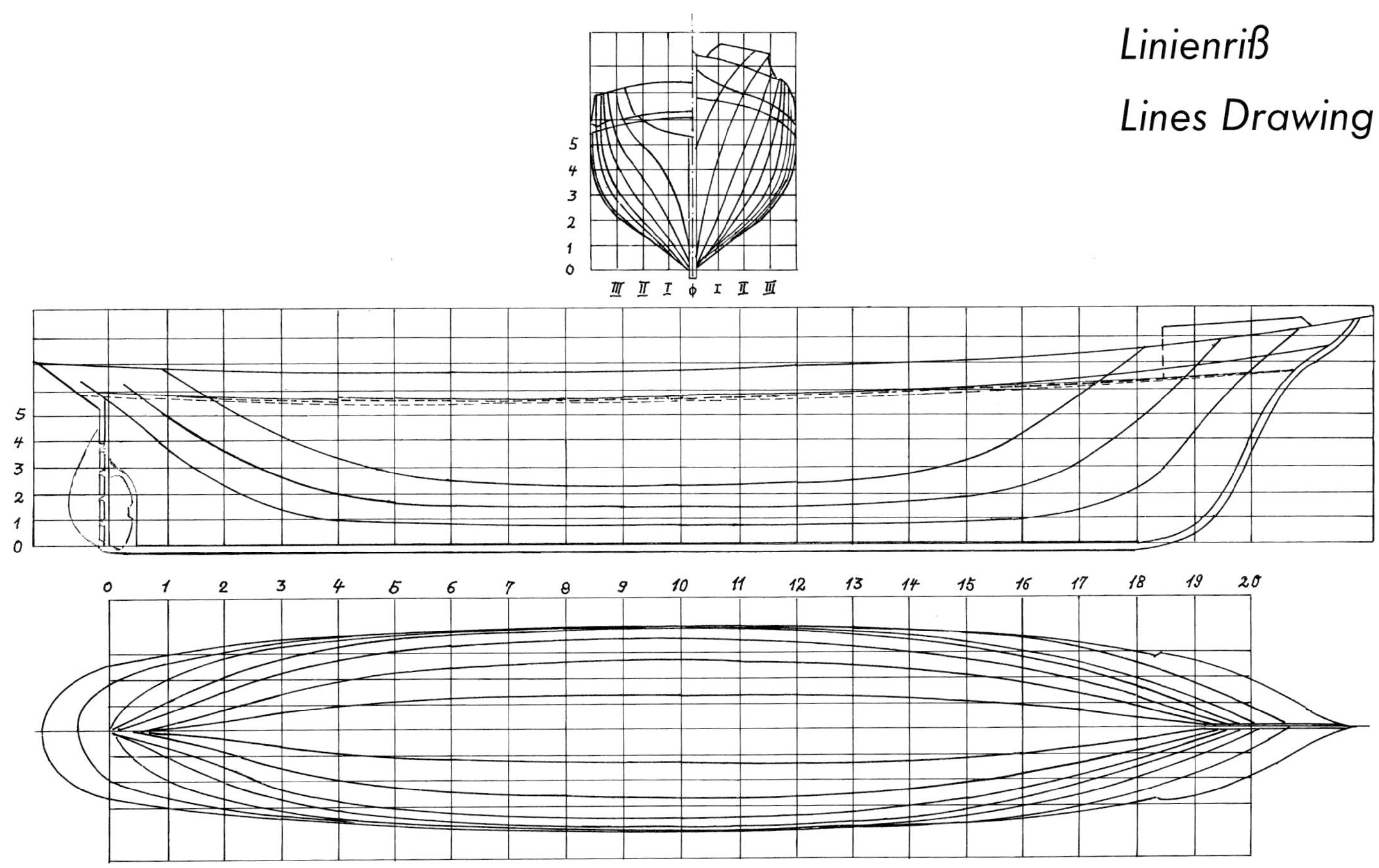

*Herausgegeben in Zusammenarbeit mit der Deutschen Stiftung Sail Training und der Sail Training Association Germany*

Folgende Unternehmen, Institutionen und Einzelpersonen haben den Umbau des Feuerschiffes „Kiel" zur Bark „Alexander von Humboldt" mit Sach-und Geldspenden oder durch anderweitige Leistungen gefördert: Bernd Ahlers, Herr Allstein von der Feuerwehr Bremerhaven, Uwe Altmann, Dieter Ansorge, Arbeitsamt Bremerhaven, Dirk Ay, Babcock Oberhausen, Dr. J. U. Bachmann, Peter Bahr, Friedrich Baier, Dieter Bankert, Cpt. Barner-Jespersen, Herr Basch, Brauerei Beck & Co., Dieter O. Beck, Herr Beckedorf, Leo R. Beckmann, Hein Behn, Firma Beken, Helmut Bellmer, Rolf Dieter Benub, Wilhelm Berding GmbH & Co., Berufsbildungs- u. Technologie-Zentrum der Kreishandwerkerschaft Bremerhaven, Werner Beutenberg, Manfred Block, Karl Blohm, Alfred Bockermann, Kay Bohlmann, Harry Böhmert, Günter Boldt, Bastian Boomert, Bernd Böttcher, Heinz Böttcher, Bürgermeister Heinz Brandt, Bremer Lagerhaus-Gesellschaft AG Bremen, Marco Brotz, Herr Brunschede, Bundesmarine, Bundesminister für Verkehr, Bundeswehrverwaltung, Dieter Büntjen, Kpt. Heinz Cepok, Zygmunt Choreń, Comet GmbH, Heidemarie Creaga, Dieter Czarnetzki, Dehnel Korrosionschutz GmbH, Bianka Dekepper, Uwe Delfs, Sabine Dierks, DLRG, Firma Döscher, Theodor Dresen, Walter Dumke, Georg Duncker, Versicherungsmakler, Max Dwillies, Erwin Ehlers, Erwin Ehlers KG, Holzhandlung Ehlers, Elektrotechnik und Aggregatebau GmbH, Jürgen Engel, Dr. Manfred Ernst, Wolfgang Felsmann, Festma Vertäugesellschaft, Harald Freiherr v. Forstner, Fa. Arthur Friederichs, Kpt. Olaf Gäde, Geestemünder Bank Bremerhaven, Erich Gensing, Herr Gerke, Bianka Gerken, Germanischer Lloyd Klassifikations-Gesellschaft, Hilke Gietz, Fa. Gleistein & Sohn, Hennig Goes, Rolf Golebski, Wolfgang Gose, F. H. Gottfeld GmbH, Andrea Grammes, Peter Grammes, Regierungsdirektor Hinrich Gravert, Arthur Gruben GmbH, Hubert Hagelstein, Axel Häger, Frank Häger, Herr Halfmann, Hermann Hanken, Hansestadt Bremisches Amt Bremerhaven, Prof. Alfred Harms, Egon H. Harms, E. H. Harms GmbH & Co. KG, J. P. Harms, Georg Harmsen, Frank Hartmann, Dr. Joachim Hartmann, HATLAPA-Uetersener Maschinenfabrik, Fa. HECO-Maschinenbau, Dipl.-Kfm. Hans J. Heider, Senatsdirektor Horst Heise, Bremen, Arnold Henkens, Klaus Herrling, Bürgermeister Paul Hertrampf, Gustav Heuer, Herr Hinrichs, Kpt. z. S. Nickel Peter Hinrichsen, Herr Hinsch, Fa. Hinsch & Ruhland, Willi Hitz, Kpt. Dipl. Ing. Günter Hoffmann, Gerd Holst, Detlev Holzhüter, Ulrike Höper, Jan Hövener, Manfred Hövener, Günther Hüneken, Herr Hupke, Fa. Imperat Farbwerk, Industrie Club, Industrie- und Handelskammer Bremerhaven, Industrie- und Luftaufnahmen Hero Lang, Erika Ivaneczko, Jan-Peter Jansen, Kpt. H.-W. Janssen, Andreas Janzik, Jugendamt Bremerhaven, Dietmar Kabisch, Peter Kaiser, Herr Kellerhoff, Gerhard Kempf, Volker Klose, Monika Kludas, Hans-Jürgen Kluge, Manfred Koch, Mario Koch, Peter Koch, Rolf Köhler, Carl Köhn, Präsident Köhn von der Wasser- u. Schiffahrtsdirektion Aurich, Bürgermeister a. D. Hans Koschnik, Helmut Kraft, Fa. Wilhelm Kräft, Tischlerei und Zimmerei Kramer, Torsten Kramer, Hans-Jürgen Krams, H. Kravagna, Dr. med. Herwig Krawen, Kreishandwerkerschaft Bremerhaven, Krupp Atlas Elektronik, Heinz Kühl, Daniel Küpfer, Norbert Kupke, Kpt. Gerhard Lickfett, Fa. von der Linden, BÜFA-Farben, Lloyd-Apotheke, Lloyd Werft, Magistrat Bremerhaven, Fa. James Marquardt, Marine-Ortungsschule, Günter Maske, Herr Mathias von der Firma Klöckner-Rodenburg, Kpt. Helmut Meesenburg, Reinhard Metz, Wilfried Meyer-Kamps, Kpt. Jochen Morgenroth, Motorenwerk Bremerhaven, Motoren- und Turbinen-Union, Nils Mühlenbrook, Robert Müller Stiftung Hamburg, Andreas Neubauer, Hafenkapitän Eberhard Nölke, Doris Norkeit, Nortrail Transport GmbH, Kpt. Klaus Novara, Herr Oldenettel, Hauke Otto, Florian Palme, Pamir-Passat-Vereinigung, Pandi Marine Insurance, Pandi Services, Dr. Michael Passow, Herr Pigors, Martin Poetting, Stefanie Poetting, D. Poggenburg, Heiko Pollin, Klaus Pollin, Kpt. z. S. Karlheinz M. Reichert, Herr v. Riegen, Fa. Rockwool, F. Rodenburg GmbH, Klaus Rodiek, Michael Rohrbach, Heinrich Rönner GmbH, Herr Rühmkorf vom Sportamt der Stadt Cuxhaven, Hans Rummel, RWO-Maschinenfabrik, Sachverständigenbüro Kpt. Bessau / Kpt. Möller, Michael Sarach, Kpt. Hartmut Schäfer, Kurt Schakie, Herr Scheppelmann, Gerold Schenk, Kai Schirmer, Museumsdirektor Gert Schlechtriem, Helmut Schmeiduch, Herr Schnelle, Hafenlotse J. Schreiber, Jürgen Schröter, Bernd Schulz, Zollamtsleiter Schulz, Britta Schwarting, Cpt. Morin Scott, Seebeckwerft, Dr. Seidemann, Shanty-Chor-Loxstedt, Kpt. D. Soitzek, Lars Spangenberg, Spitzkowsky GmbH & Co. KG, Kpt. z. S. Hans Freiherr v. Stackelberg, Städt. Sparkasse Bremerhaven, Frank Steinhauer, Lotsenältermann Steinhauer, Herr Stender, Martin Strampp, Tecklenborg, Kegel & Co., Kpt. Prof. Dr. Ing. Karl Terheyden, Otto Terhorn, Egon Thiel, Egon Thomsen, Herr Töns von der Firma Käfer Isoliertechnik, Trans-Ocean, Verein zur Förderung des Hochseesegelns, Transport und Services GmbH, Präsident Treu von der Oberfinanzdirektion Bremen, Renate Tscherner, Herr Tymian, Kpt. Wolfgang Vollmer, Fa. Adolf Voss & Co., Dr. Udo Walger, Frank Waltel, Wasserschutzpolizei Bremerhaven, Wasser- und Schiffahrtsdirektion Aurich, Wasser- u. Schiffahrtsdirektion Wilhelmshaven, Firma H. Wegener & Sohn, Wolfgang Westphal, Dr. Wolfgang Wiedemann, Horst Wiesner, Hartwig Wiest, Admiral a. D. Horst Helmut Wind, Kpt. Rudolf Wittenhagen, Heinrich Woermann, Herr Woetzel, Ronald Wöhrn, Bernd Wormland, Ando Yoo, Uwe E. Zoller GmbH. Sollte einer der vielen Förderer ungenannt geblieben sein, bittet der Verlag um Nachsicht.

Für freundliche Hinweise und Hilfe beim Zustandekommen dieses Buches sind Autoren und Verlag zu Dank verpflichtet Kpt. Rainer von Ahlften, Peter Althini, Dipl. Ing. Friedrich Baier, Kpt. Günter Barty, Leo R. Beckmann, Baudirektor Bedke, Alfred Bockermann, Harry Borenius, Janet E. Bradbury, Dipl. Ing. Jobst Broelmann, Kpt. Fritz Brustal-Naval, Zygmunt Choreń, Kpt. Robert Drögemöller, Egon Eggerstedt, Kjell Ekström, Kpt. Uwe Fischer, Kpt. Axel Olaf Gäde, Hennig Goes, Günter Grätsch, John H. Hamilton, Harald Hans, „Hansa"-Archiv, Prof. Dipl. Ing. Alfred Harms, Reeder Egon H. Harms, Hans J. Heider, Kpt. Helms, Gerhard Hildenbrand, Dr. Hofmeister, Herr Hoisen, Frau Höppe, Kpt. Manfred Hövener, Herr Hüsmann, Frau Ilgauds, Institut für Schiffbau der Universität Hamburg, Werner Istel, Museumsdirektor Dr. Jensen, Kpt. Jürgen von Karmainsky, Arnold Kludas, Manfred Krüger, Prof. Dr. Peter Kuckuck, Baudirektor Siegfried von Lilienfeld-Toal, Lotsenbrüderschaft am Nord-Ostseekanal, Nils O. Lundberg, Klaus-D. Menke, Kpt. Klaus Novara, Herr Pröwrock, Dr. Roder, Klaus Rodiek, Kpt. Friedrich Karl Rühr, Eberhard Schmiel, Dipl. Ing. Bernd Schulz, Christian Söderström, Siv Strandberg, Tor Strandberg, Kpt. Göte Sundberg, Rudolf Tscharntke, Baudirektor Volkhard Wetzel und Kpt. Rudolf Wittenhagen.


Satz: Lichtsatz Imken, D 2900 Oldenburg. Reproduktionen: Litho Niemann, D 2900 Oldenburg. Druck und Einband: Druckhaus Neue Stalling GmbH & Co KG, D 2900 Oldenburg.

ISBN 3-924896-13-5

Die Aufnahmen für das Buch, eine Auswahl aus mehr als sechstausend Einzelphotos, stammen von: Günter Barty (Seite 9); Leo R. Beckmann (Seiten 10, 16 links und rechts, 17); Herbert H. Böhm (Seite 64); Axel Olaf Gäde (Seiten 18 links, 21, 23); Andrea Grammes (Seite 22 links und rechts); Kai Greiser (Seiten 35, 56 / 57, 59, 98); Clas Broder Hansen (Seiten 24, 25, 31, 34 links und rechts, 36 links und rechts, 38, 43, 44, 45, 47, 48 links und rechts, 49, 50, 52 links und rechts oben, links und rechts unten, 54, 55, 60, 61, 62, 63, 68, 69 links oben, 72, 73, 74 oben und unten, 75, 80 links und rechts, 82 / 83, 84, 85, 86, 87, 88 / 89, 94 rechts oben, 96, 99 rechts oben und unten, 100 / 101, 103, 106, 107); Hans Jürgen Hansen (Seiten 2 / 3, 37, 42, 53, 69 rechts oben und unten, 76, 105); Monika Kludas (Seiten 12, 15, 18 rechts, 20, 40, 41, 51, 58, 65, 66, 67, 69 links unten, 70 / 71, 77, 78, 79, 81, 90, 91, 92 links und rechts, 93, 94 links oben und unten und rechts unten, 95, 99 links oben und links unten, 102); Hero Lang (Seiten 4, 13, 19, 26 links, Mitte und rechts, 27, 28 links und rechts unten, 29, 30, 32 / 33, 39, 97); Bodo Mueller (Seite 104); Renard (Seite 8); Bernd Schulz (Seite 46) und Ino Steffen (Seiten 11, 14).